Structural Analysis

A Historical Approach

T0276020

This volume provides a concise, historical review of the methods of structural analysis and design – from Galileo in the seventeenth century, to the present day. Through it, students in structural engineering and professional engineers will gain a deeper understanding of the theory behind the modern software packages they use daily in structural design. This book also offers the reader a lucid examination of the process of structural analysis and how it relates to modern design.

The first three chapters cover questions about the strength of materials, and how to calculate local effects. An account is then given of the development of the equations of elastic flexure and buckling, followed by a separate chapter on masonry arches. Three chapters on the overall behaviour of elastic structures lead to a discussion of plastic behaviour, and a final chapter indicates that there are still problems needing solution.

Structural Analysis
A Historical Approach

Jacques Heyman

Emeritus Professor of Engineering,
University of Cambridge

CAMBRIDGE
UNIVERSITY PRESS

CAMBRIDGE UNIVERSITY PRESS
Cambridge, New York, Melbourne, Madrid, Cape Town, Singapore, São Paulo

Cambridge University Press
The Edinburgh Building, Cambridge CB2 8RU, UK

Published in the United States of America by Cambridge University Press, New York

www.cambridge.org
Information on this title: www.cambridge.org/9780521622493

First published 1998
This digitally printed version 2007

A catalogue record for this publication is available from the British Library

Library of Congress Cataloguing in Publication data

Heyman, Jacques.
Structural analysis : a historical approach / Jacques Heyman.
p.   cm.
Includes bibliographical references and index.
ISBN 0 521 62249 2
1. Structural analysis (Engineering)–History.   I. Title.
TA645.H49   1998
624.1′71′09–dc21   97-18474   CIP

ISBN 978-0-521-62249-3 hardback
ISBN 978-0-521-04135-5 paperback

Contents

Contents

Preface

The business of the structural engineer is to make a design to meet some specified brief – for example, a steel-framed factory to house a manufacturing activity; a bridge to span a wide estuary; a gantry to carry an overhead cable for an electric train. Design criteria must first be identified – heavy crane loads may be critical for the factory, wind-induced vibrations for a suspension bridge, accurate location of the cable for the train. To satisfy these criteria the engineer makes calculations, and it has proved convenient, during the last century and a half, whether explicitly recognized or not, to divide the engineer's activity into two parts.

In the first stage, *The Theory of Structures* is used in order to determine the way in which a structure actually carries its loads. There are many alternative load paths for a (hyperstatic) structure; one of these will be chosen by the structure, and must be discovered by the engineer. This formulation of the problem seems to imply something more than a dispassionate search for truth; the structure seems somehow to have anthropomorphic qualities, and indeed nineteenth (and twentieth) century notions such as those of 'least work' may colour the engineer's judgement. For example, in forming such ideas the designer may assume unthinkingly that of course there *is* an actual state of the structure in which it will be comfortable. It is, however, a matter of fact that the structural equations are extremely sensitive to very small variations in the information used in any structural analysis, and that the structural action, the state in which a structure finds itself to be comfortable, can show enormous variation caused by trivial imperfections in manufacture or construction.

Be this as it may, when the engineer has determined, according to his theories, a particular set of primary structural forces, then the second stage of structural design is reached. The scene is shifted from an overall view of the structure to a detailed examination of response – for example, can the force calculated by the designer to act at a certain cross-section be sustained safely by the material present at that section? This examination is carried out within the framework of what is sometimes known as *The Strength of Materials*, and, conventionally, involves the calculation of (elastic) stresses throughout the structure.

The two stages, theory of structures and strength of materials, are logically and actually distinct, but cannot be separated in practice in the elastic design of a hyperstatic structure. Section properties must be assigned before primary forces can be determined; once those forces are known, section properties can be assigned. The elastic design process is thus circular; calculations must proceed by trial and error, by iteration, or by following some schema known to lead to a satisfactory design. Stresses may then be checked against given criteria; indeed until the twentieth-century notion that 'plastic' information might be valuable in approaching the design of structures, the calculation of elastic stresses has been the main activity of the designer for the last 170 years (from, say, 1826). *The Theory of Elasticity* deals with this branch of the science, and gives rise to elegant and complex mathematics.

This mathematics is massively (if not exhaustively) explored in the well over 2000 pages of Todhunter and Pearson's *(A history of) The theory of elasticity (and of the strength of materials) (from Galilei to Lord Kelvin)*. Prominence is given on the title page to *The Theory of Elasticity*, and the three volumes are described as being by the late Isaac Todhunter, edited and completed by Karl Pearson. (Todhunter had died in 1884, and the volumes were published in 1886 and 1893.) Todhunter was a distinguished Cambridge mathematician, and Pearson was a professor of applied mathematics at University College, London.

Truesdell, in his magnificent study of *The rational mechanics of flexible or elastic bodies 1638–1788*, which forms his introduction to two volumes of *Leonhardi Euleri Opera Omnia*, cites the work of Todhunter and Pearson, and identifies Pearson's particular contributions. Truesdell takes many opportunities to denigrate Pearson's understanding of the subject; he is particularly upset that Pearson had supplied a frontispiece showing drawings of 'Rupture-Surfaces of Cast-Iron' to a work whose title restricts it to *theory*.

There is no doubt that both Todhunter and Pearson (and Truesdell) were accomplished mathematicians; what is at question here is how mathematics should be applied to the study of structures. Whatever their actual understanding may have been, it is tempting to classify Truesdell (and perhaps Todhunter) as pure mathematicians (or 'scientists'), and Pearson as an applied mathematician (or 'engineer'). Scientists and engineers speak the same language and use the same tools, and their activities are sometimes confused. There is, however, a clear distinction; the scientist uses the body of scientific knowledge to deepen the understanding of a particular science, while the engineer, with the same body of knowledge, uses it in order to do something. In the case of the structural engineer, the theory of structures, the strength of materials and the theory of elasticity form the body of knowledge. Of these, the theory of elasticity is the most interesting mathematically, and Truesdell seems content to pursue it for its own sake. Pearson, in giving pride of place to drawings of rupture surfaces, showed that he was aware of the primary objective of his particular science, which is to actually design, and analyse, a real construction.

The whole matter is, naturally, not so simple as might be implied by the last paragraph. In the first place, the engineer may find that the present state of the body of structural knowledge is inadequate to solve a particular problem; the engineer must then turn scientist, and attempt as engineer to widen scientific understanding. Secondly, the pursuit of a mathematical problem, without heed to its practical consequences, may lead to results which are in fact of immediate application. An outstanding example is that of the shape of the elastica; this problem was posed by Daniel Bernoulli as an interesting piece of mathematics involving the calculus of variations, and it was solved brilliantly by Euler, as a mathematical exercise, in 1744. Euler himself, however, understood that the basic result could be applied to the buckling of columns; the work is described in Chapter 4.

This book, then, as implied by its title, is concerned with the theory of structures as applied to the real problems of structural analysis and design. Although it mentions building codes, it does not deal with their development; codes are very necessary to guide practical designers, but they reduce the theory to sets of rules which can be applied without enquiry, without thought even, into the basis for those rules. (Such codes have existed for well over two millennia, and the short introduction to Chapter 1 gives glimpses of a history that is in itself a fascinating study.) Rather, this book traces the development of theory relevant to structural design, whether this was initiated by an engineer seeking a solution to a particular problem, or was directed by a scientist seeking truth in a more abstract way. The history does not attempt to cover more than the significant developments – ignoring tempting but misleading byways, it traces the paths taken by reason through the dark foundations of the subject (as Coulomb might have put it) groping towards the truth as it is now seen.

1

Galileo's Problem

Stresses in ancient and medieval structures are low. The stone in a Greek temple, in a Gothic cathedral, or in the arch ring of a masonry bridge, is working at a level one or two orders of magnitude below its crushing strength. This is a necessary condition for survival through the centuries; it is not sufficient. It is necessary also that the *shape* of the structure should be correct, so that structural forces may somehow be accommodated satisfactorily; this is a question of correct geometry. Thus for such structures the calculation of stress is of secondary interest; it is the shape of the structure that governs its stability. All surviving ancient and medieval writings on buildings are concerned precisely with geometrical rules. The architects had, no doubt, an intuitive understanding of forces and resulting stresses, but this understanding was not articulated in a form that would be of use in design; there is no trace in the records, over the two or three millennia for which they exist, of any ideas of this sort.

Instead, the design process would have proceeded by trial and error, by recording past experience, by venturing, more or less timidly, into the unknown, and by the use of models. A large-scale model served several functions – to demonstrate the design to the commissioner, for example, and to solve constructional problems; above all, if the model were stable, so would be the full-scale building, since the model proved that the geometry was correct. All of this experience was recorded, and refined into rules of construction. In modern terms, a building science was established, and the rules expressed the theory behind the practice.

Such recording can be done verbally or by drawing. Chapters 40, 41 and 42 of Ezekiel, for example, record at interminable length the sizes of gateways, courts, vestibules, cells, pilasters and so on, for a great temple; part of a building manual of about 600 BC seems to be bound in with the books of the Old Testament. Of great interest is Ezekiel 40:3 and 5: 'I

1

saw a man ... holding a cord of linen thread and a measuring rod ... The length of the rod ... was six cubits, reckoning by the long cubit'. The long cubit was about 20.7 inches or 525 mm, so that the measuring rod was something over 3 m in length. The rod was the 'great measure', without which (in the absence of standard units) work could not proceed on an ancient or medieval building site. The cubits were subdivided into palms, and the great measure could therefore be used to establish the major dimensions of rooms as well as small individual dimensions, merely by using the numbers so diligently listed in the chapters of Ezekiel.

All of this would have been immediately recognized by Vitruvius (c. 30 BC), writing five centuries later; Frankl's (1960) exegesis makes it clear that the *ordinatio* of Vitruvius is nothing other than the great measure. As an example, Vitruvius gives proportions for the construction of temples, in which the diameter of a column is taken as the module – for the eustyle (one of the five standard arrangements) the distance between columns, the intercolumniation, should be two and a quarter modules, except that the central spacing should be three modules. The columns should be of length nine and a half modules, their bases of thickness half a module, and so on. Such rules, generated from experience, have ensured that Greek and Roman temples have survived.

Vitruvius was not lost in the 'dark ages'; his book was copied again and again for use in monastic schools and in the masonic lodges. The 'secrets' of the lodges were numerical, and recognizably Vitruvian; rules of proportion were at the heart of the Gothic building. For the construction of Milan Cathedral, for example, over a hundred years after the end of the High Gothic period, the original great measure was of 8 braccia (the braccio (arm), the Milanese 'cubit', is about 600 mm). Numerical problems with the great measure led to an expertise in 1392, but the work was totally stopped in 1399, when experts came from all over Europe to decide how construction should proceed (Ackermann (1949)).

Mignot, a master from Paris, drew up a list of 54 points in which he found the work at Milan to be defective. There are structural objections – buttressing is insufficient; and geometrical objections – bases of piers are not in the right proportions. Mignot seems clearly to have been checking the work at Milan against his own lodge's building manual; the Italian masters had no reasoned replies, and fell back on the accusation: *scientia est unum et ars est aliud.* By *scientia* they meant the theory embodied in the design rules of the manual, and by *ars* they meant the mason's art, the practice of construction. That is, the Italians acknowledged that Mignot might have had a fine set of rules, but they knew, in practice, how to build a cathedral. To which Mignot replied: *ars sine scientia nihil est*; practice is nothing without theory.

This looks to be the first sign of a rational approach to the science of building. In fact, Mignot had no deep understanding of his manual. His theory was the distillation of the design rules of Greek, Roman and medieval architects, probably codified one or two centuries earlier in the middle of the High Gothic period. They were geometrical rules found to be effective for buildings whose materials worked at low stresses. If a building were satisfactory, then it would be satisfactory when built at twice the scale.

1.1 The Dialogues

Right at the start of his *Discorsi*, Galileo strikes at the heart of this medieval theory of structural design. Salviati speaks: 'Therefore, Sagredo, give up this opinion which you have held, perhaps along with many other people who have studied mechanics, that machines and structures composed of the same materials and having exactly the same proportions among their parts must be equally (or rather, proportionally) disposed to resist (or yield to) external forces and blows. For it can be demonstrated geometrically that the larger ones are always proportionally less resistant than the smaller.'

Galileo's *Dialogues concerning two new sciences* were published in Leyden in 1638, when Galileo was 74. Five years earlier he had been convicted of heresy, sentenced to life imprisonment and forbidden to publish any more books on any subject. Holland was, of course, outside the reach of the Inquisition, and the Elseviers agreed to publish Galileo's manuscript.

The work is in the form of four dialogues (a fifth dialogue, on percussive force, was added to the second, posthumous, edition of 1644). The three Interlocutors are Salviati, who speaks for Galileo; Sagredo, who represents Galileo as a younger man, and puts forward views on occasion that the older Galileo has rejected; and Simplicio, who might represent a very young Galileo, and who acts as a foil to the other two more learned scientists. Each dialogue is supposed to span a day. The third and fourth deal with the development of a science of (pre-Newtonian) mechanics; it is the second day's dialogue that is concerned mainly with structural matters.

However, Sagredo's 'brain is already reeling' at the start of the first day when the mature Galileo, Salviati, mounts his attack on medieval geometry. He has much more to say about the 'square/cube law' on day

2 of the *Dialogues*, but he starts immediately by introducing an example that is, effectively, the archetypal structural problem with which Galileo is concerned.

Salviati imagines a horizontal wooden pole of given dimensions with one end fitted into a vertical wall; the pole thus acts as a cantilever beam. If the length of the pole is increased there will come a point at which it breaks under its own weight; a shorter beam could carry additional load, and longer beams would break at once. It is the breaking of cantilever beams that forms the main subject of Galileo's second day of the *Dialogues*. As will be seen, he makes a structural analysis to determine the value of the greatest bending moment in the beam, and he equates this to the moment of resistance of the cross-section. Thus a calculation of 'theory of structures' is combined with one of 'strength of materials' in order to solve the problem.

Immediately after the introduction, however, on the first day, Galileo has a seeming digression concerned purely with the theory of structures. Salviati tells a story concerning a very large marble column that was stored horizontally, the weight being supported on two baulks of timber near its ends. It occurred to a workman that the column might break under its own weight at the middle, so he inserted a third similar baulk of timber there. After a few months the column was found to be broken precisely over the third inserted support. Salviati explains how this had come about. It was found that one of the baulks near the end of the column had rotted and settled, while the inserted central support remained sound; effectively, then, one half of the column was unsupported. Had the column remained supported only by the original two baulks, all would have been well; if one baulk had settled, then the column would merely have followed.

This glimpse of the possible behaviour of statically indeterminate structural systems is not further discussed by Galileo. However, when, towards the end of the second day, he has solved to his satisfaction the problem of the breaking of a cantilever, he applies his results to the failure of a simply supported beam (the marble column on its end baulks) and of a beam supported on a central support.

Galileo has already stated clearly (although indeed only implicitly) in the first few pages of the first dialogue that his subject is the *fracture* of beams, and he next makes an enquiry into what is happening when a piece of wood, or any other solid, breaks. To clarify the discussion (as he says), he imagines a cylindrical specimen (of wood, or other material; his accompanying illustration, fig. 1.1, appears to be of a stone pier) to

Fig. 1.1. Galileo's imaginary tension test.

be hanging vertically, and loaded by an increasing weight at the bottom until it breaks, 'just like a rope'. Thus starts a series of digressions which make up the substance of the whole of the first day's dialogue; Simplicio can imagine that the longitudinal fibres (*filamenti*) extending the whole length of a wooden specimen can make the whole specimen strong, but asks how a rope, composed of fibres only two or three braccia long, can be equally strong. Salviati explains how these short fibres are twisted together to form a long rope, their mutual interaction conferring strength on the whole.

However, it is to the fracture of apparently amorphous materials, such as marble, metal or glass, that the discussion now turns. Salviati admits the difficulty of this problem, and has the view that the particles of a body have some inherent tenacity, and he also refers to the well-known repugnance that nature exhibits towards a vacuum. The question of voids arises from the experience that two slabs of material, if smoothed, cleaned and polished, may be slid one against another but are difficult to pull apart in tension. From this observation is developed the idea that an apparent solid might consist of a very large number of small parts having inherent strength, and also of a very large number of voids, both contributing to the overall strength. Indeed, the number of voids could be infinite, and many pages of the dialogue (say an hour or so of discussion

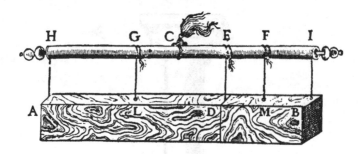

Fig. 1.2. Galileo's diagram explaining the law of the lever.

on the first day) are devoted to mathematical demonstrations concerned with this concept – that a finite area, for example, can contain an infinite number of voids.

One topic leads naturally to another, and Salviati is constantly reminded of interesting results from geometry, to the point where Sagredo remarks on how far they have strayed from their subject. But Salviati has been concerned with trying to explain how expansion and contraction (due to temperature change, for example) can occur without assuming interpenetration of bodies or the introduction of voids and, almost inevitably, discussion turns to Aristotelian theories of motion.

Motion is the second new science, and is exposed on the third day of the dialogues, but it is on the first day that Galileo makes the famous statement that two bodies of different weights will fall under gravity at the same speeds. Simplicio finds this hard to believe, although Sagredo says he has made the test (it is thought unlikely that Galileo dropped weights from the leaning Tower of Pisa). It is clear that the test had been made somewhere; Salviati, explaining to Simplicio, notes that Aristotle says: 'A hundred-pound ball falling from the height of a hundred braccia hits the ground before one of just one pound has descended a single braccio.' In fact they arrive at the same time, or rather (and this is the evidence of a test), the larger is ahead by two inches, and Salviati asks Simplicio not to hide behind those two inches the ninety-nine braccia of Aristotle. There follows then a long discussion of motion in dense and rarefied media, water and air; and then of the vibration of the pendulum; and so to the vibration of musical strings. Even Salviati, at the end of the first day, wonders how they have allowed the discussions to be carried on through so many hours, without tackling the main problem.

This problem is stated again clearly at the start of the second day: it is to find the strength of a bar when it is broken as a cantilever. The breaking force is in fact a 'bending moment', an idea which relates to Aristotle's analysis of the lever (but which, says Salviati, was handled better by Archimedes). Salviati has to expound the theory of moments, and, to do this, Galileo introduces an idea which is, in essence, that of the 'free-body diagram'. A (uniform) baulk of timber AB, fig. 1.2, is hung by two strings HA and IB from a rod HI (this rod is referred to later as a 'balance' (*balancia*) or 'steelyard' (*libra*)). Evidently a string at C at the midpoint of HI will carry the baulk AB in equilibrium. The baulk is then imagined to be cut into unequal parts at D, and a third string ED is introduced, attached at D both to the portion DA and the portion DB of the cut baulk. Equilibrium will not be disturbed. Two further strings are now introduced, at G and F, over the midpoints L and M of the portions of the baulk, and the three original strings are removed, again without upsetting equilibrium. Thus two heavy bodies, DA and DB, can be supported by two strings GL and FM from a steelyard, the steelyard itself being supported at C. Galileo shows easily that the ratio GC to CF is as the weight of DB to that of DA. (Although algebraic equations were acquiring modern form in Galileo's time, he was using the classical theory of ratios to be found in Euclid.)

After some further exploration of the lever (in the form of a crowbar to lift weights from the ground) Galileo is finally ready to tackle the problem; fig. 1.3 shows his illustration which has become a sort of icon for the application of rational mechanics to the theory of structures. The member ABCD is prismatic, and made of glass, steel, wood or any other frangible material; when loaded at C, it is evident that if it breaks, it will break at B where it is embedded in the wall. Now the member has a certain 'absolute strength' (that is, the tensile strength to which attention was devoted on the first day); the moment of this absolute strength about B must equal at fracture the moment of the applied weight about B. Galileo is considering the moment equilibrium of a cranked lever, fig. 1.4 (he does not draw this figure); with modern notation,

$$W\ell = \frac{1}{2}Sd, \qquad (1.1)$$

where S is the absolute (i.e. tensile) strength of the member, which has breadth b and depth d. (If the beam has a circular cylindrical cross-section, then Galileo notes that the lever arm for the absolute resistance is the radius r.)

In terms of modern ideas of stress, Galileo has evidently assumed

Fig. 1.3. Galileo's cantilever beam.

a uniform distribution of limiting stress σ_0 at fracture, as indicated in fig. 1.5. Thus, for the rectangular section, the value of S is $bd\sigma_0$, and the section modulus implicit in the right-hand side of equation (1.1) is $\frac{1}{2}bd^2$.

As is evident, figs 1.4 and 1.5 (which, it must be repeated, were not drawn by Galileo) are incomplete free-body diagrams – the reactions at the fulcrum B are not shown. Seventeenth and eighteenth century work was concerned to 'correct' Galileo's analysis in this respect; the form of equation (1.1) is unchanged, but the factor of $\frac{1}{2}$ was determined to have other values by different writers. Galileo himself, however, does not make use of the numerical value of absolute strength S; he is concerned with calculations of relative strengths, and the quantity $\frac{1}{2}S$ could be looked on as a given physical parameter entering the analysis. For example, Galileo shows easily and correctly that the *ratio* of loads T to X required to break a rectangular-section cantilever beam when it is on edge and when it is flat is simply the ratio ca/cb, fig. 1.6; since the section modulus has

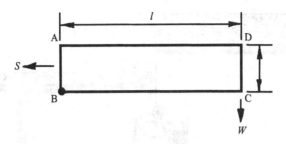

Fig. 1.4. Conditions at fracture of a cantilever beam.

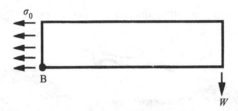

Fig. 1.5. Stress distribution at fracture implicit in Galileo's analysis.

been determined to be of the correct form (kbd^2), the factor ($k =)\frac{1}{2}$ in equation (1.1) is irrelevant to the value of this ratio.

Galileo then discusses problems of other ratios of strengths – for example, he shows that the strength of a circular cylindrical cantilever is proportional to the cube of the diameter. Similarly, a beam breaking under the action only of its own weight has a strength proportional to the square of its length. Other examples follow, all effectively demonstrating the square/cube law, culminating in Salviati's general statement about the impossibility of building enormous ships, palaces and temples – 'nor could nature make trees of immeasurable size, because their branches would eventually fail of their own weight'. Figure 1.7 shows the shapes of two bones, one three times the other in length, and then thickened so that it could function in a large animal as the smaller bone performs for the small animal.

The breaking of a beam has been analysed with respect to a cantilever, fig. 1.3. Galileo identifies the upper beam in fig. 1.8 as two cantilevers back to back, and states correctly, but not clearly, that the lower simply supported beam DEF will break under its own weight when the length is twice that of the corresponding cantilever of critical length. (These

Fig. 1.6. Comparative strengths of a wooden beam on edge and flat.

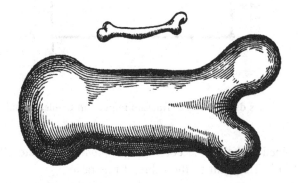

Fig. 1.7. Two bones having similar functions; the large animal is three times the (linear) size of the smaller.

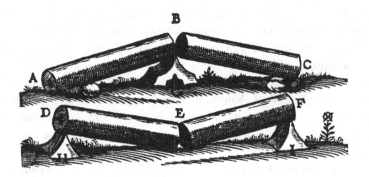

Fig. 1.8. The breaking of a beam supported at its centre, and at its ends.

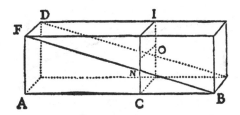

Fig. 1.9. A cantilever of varying cross-section.

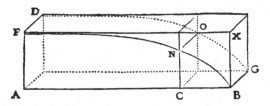

Fig. 1.10. A cantilever of 'equal resistance' – the curve is badly drawn.

examples refer, of course, to the stored marble column mentioned on the first day.) Further, he then tackles clearly the problem of a stick held in the hands and broken over the knee; the knee should be placed at the centre of the stick. Finally, in this small digression he finds in effect the maximum bending moment in a simply supported beam when a transverse point load is applied at a general point in the span.

Almost the last problem to be tackled on day 2 (before moving to study motion on days 3 and 4) is that of the correct shape of the cantilever beam of equal resistance, that is, one that would fracture in bending simultaneously at each of its cross-sections when loaded by a single tip load. A full prismatic cantilever beam, fig. 1.9, loaded at its tip B, will fail only at the root AF. If the same beam is sawn along the diagonal FB, then the lower tapering beam FAB under its tip loading is weaker under the bending moment acting at the general section CN than it is under the bending moment acting at the root AF; the beam FAB cannot therefore carry any load that would cause fracture at the root. There must be some cut that can be made to leave a solid of equal strength at each cross-section. Galileo shows that the beam, of constant width, should have a depth that varies as the square of the distance from the tip, that is, as stated by Salviati, a parabola with vertex at B. The illustrative sketch, fig. 1.10, is badly drawn; the curve seems to have the vertex at

Galileo's Problem

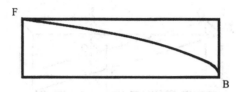

Fig. 1.11. Correct shape of cantilever of equal resistance.

F, and corresponds to the path of a projectile launched horizontally at F. The correct shape is that sketched in fig. 1.11. (As has been noted, Galileo was not in control of the publication of his book; he had in any case become blind before he received a printed copy.)

So ends the exposition of the first of Galileo's two new sciences. The mechanics of motion discussed on the second pair of days is perhaps more formal in presentation, and contains pedagogic material in Latin inserted in the Italian dialogues.

Todhunter and Pearson note that the 'problem of solids of equal resistance led to a memorable controversy in the scientific world'.

2

The Moment of Resistance

As has been seen, Galileo's problem was the determination of the ultimate moment of resistance of a member (wooden, stone, metal, glass) in bending. The problem was posed by reference to a cantilever beam, acted upon by a tip load, or its self-weight, or both; the value of the breaking load(s) was sought. Static equilibrium requires that the moment of the applied load(s) at the root of the cantilever must equal the moment of resistance of the cross-section; since the problem is statically determinate, a problem in the theory of structures is transformed into a problem of strength of materials. Galileo, and later scientists, did not of course think in this way; in particular, the notion of hyperstatic structures, for example the beam on three supports or, later, the propped cantilever or the fixed-ended beam, is not made explicit. These last two more complex structures were in fact discussed in 1798 by Girard, and 'correct' solutions were found (see Chapter 6); the solutions were, however, specific for the problems, and Girard does not make general statements about statical indeterminacy. Such ideas became formalized a quarter of a century later; the date of 1826, when Navier published his *Leçons*, is a convenient marker, and indeed it was not until over a century after that date that the straitjacket imposed by Navier on structural design was finally loosened.

Girard starts the introduction to his book (on the strength of materials and on solids of equal resistance) by stating that his subject consists of something more than rigid-body statics. Nature has not created perfectly rigid materials, says Girard, and he foreshadows the modern general statement that the theory of structures is concerned with the mechanics of slightly deformable bodies. He gives a fairly comprehensive history of the development of the theory of bending for the century and a half after Galileo, and it is apparent that, very soon after publication of the *Two new sciences*, those interested in the problem were questioning Galileo's

13

assumption of 'inextensibility'. Galileo had supposed the beam to be composed of parallel fibres along its length; at fracture at the root of the cantilever each of these fibres would reach its ultimate strength as given by the tension test. Galileo apparently carried out no experiments to verify, numerically, his results (as was seen in Chapter 1, Galileo had obtained the correct *form* for the expression for the moment of resistance of a cross-section – this matter is discussed further below). The first such experiments were apparently made in Sweden by P. Wurtz; Girard mentions a letter of 1657 from François Blondel to Wurtz referring to the matter, but Saint-Venant (1864) in his edition of Navier's *Leçons* could find no details. Wurtz was certainly interested in the question of solids of equal resistance, broached by Galileo towards the end of the second day; a book on the subject was published by Marchetti (1669), and heated discussion (discussed briefly by Girard and mentioned by Todhunter and Pearson) culminated in two publications by Guido Grandi (1712).

The search for the solution to the problem of solids of equal resistance seems now to represent a side path in the development of structural engineering. It was clearly attractive to scientists, involving as it did exploration of territory that was just about to be opened by the invention of the calculus. Further, however unreal and impractical a beam of parabolic profile might appear, such a 'design' would give some sort of basis against which any other solution could be matched.

2.1 Mariotte 1686

Galileo's main problem, however, continued to be treated, and Mariotte made a major contribution in 1686. He was concerned, in part 5 section 2 of his *Traité du mouvement des eaux*, with the strength of water pipes, and he derived a correct expression for the wall thickness of a pipe of given diameter to support a given pressure. In the course of his work Mariotte made both tension and bending tests on his materials, and he could not relate the results of the two by Galileo's formula. He concluded that Galileo's assumption of inextensibility was incorrect; rather even the hardest materials (and he had tested glass and marble as well as wood) must show some extension when loaded. He assumed, in effect, that behaviour was linear-elastic, and identical in tension and compression; further, there was a maximum extension that could be imposed on a given material, and fracture would occur if that limit were passed.

Using these ideas, and retaining Galileo's position of the 'neutral axis' at the base of the section, Mariotte deduced that the stresses should be

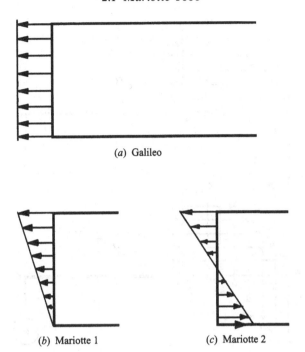

(a) Galileo

(b) Mariotte 1   (c) Mariotte 2

Fig. 2.1. Stress distributions at root of cantilever.

distributed to give a triangular block, Mariotte 1 in fig. 2.1. The use
of the word 'stress' is anachronistic; a better idea of Mariotte's way of
thinking, of loaded parallel 'fibres', is given by Girard's illustrations of
bent levers, fig. 2.2. In this figure the upper illustration, with equal fibre
'weights', corresponds to Galileo's solution, the lower to Mariotte's. The
corresponding section modulus of a rectangular beam, $b \times d$, is $\frac{1}{3}bd^2$,
compared with the $\frac{1}{2}bd^2$ of Galileo.

Mariotte immediately abandons this analysis, stating that it may be
imagined that, while the upper fibres of the cantilever beam are extended,
those on the lower face of the beam are compressed. He places the neutral
axis at mid-height of the section, Mariotte 2 in fig. 2.1; he gives no proof,
but he has stumbled on the correct solution to the problem of elastic
bending. However, he does not use such words; Mariotte was concerned
with the problem of the breaking load of a cantilever beam. In working
out the value of the section modulus corresponding to fig. 2.1(c), by a
singular inadvertence (as remarked by Saint-Venant) Mariotte dropped

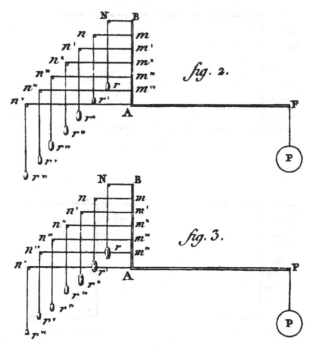

Fig. 2.2. Girard's figs 2 and 3. Figure 2 corresponds to Galileo's solution (fig. 2.1(*a*)), and fig. 3 to Mariotte's (fig. 2.1(*b*)).

a factor of 2, and obtained his previous result of $\frac{1}{3}bd^2$ instead of the correct value of $\frac{1}{6}bd^2$.

In fact, Mariotte followed Galileo in making his calculations in terms of relative and absolute strength rather than in terms of section modulus. With modern notation for a material having a limiting stress σ_0, the absolute strength S for a rectangular cross-section (the maximum tensile force that can be applied) is $S = bd\sigma_0$. The relative strength in bending, that is, the moment of resistance, is obtained by multiplying S by an appropriate fraction of the depth of the section. Thus Galileo had derived the relative strength as $\frac{1}{2}Sd$, and Mariotte as $\frac{1}{3}Sd$; on the same basis, the correct value should be $\frac{1}{6}Sd \left(= \frac{1}{6}bd^2\sigma_0 \right)$.

The tests made by Mariotte were on $\frac{1}{4}$ inch diameter cylindrical specimens of dry wood; he determined the absolute strength from a tension test as $S = 330$ lb. The circular rods were then tested as cantilevers of length 4 in, and the results were interpreted by applying, without comment, the theory for rectangular beams. (The distribution of fig. 2.1(*c*),

Table 2.1. *Strength of a cantilever beam*

$\left(S = 330 \text{ lb}; d = \frac{1}{4} \text{ in}\right)$	Relative strength (in lb)	Breaking load (lb)
Galileo $\left(\frac{1}{2}Sd\right)$	41.25	10.3
Mariotte $\left(\frac{1}{3}Sd\right)$	27.5	6.9
'Correct' $\left(\frac{1}{8}Sd\right)$	10.31	2.6
Observed	(24)	6

if applied to a beam of circular cross-section, gives a relative strength of $\frac{1}{8}Sd$.) Numerical values are displayed in Table 2.1. Had he not made his mistake with the factor of 2, Mariotte's theoretical breaking load would have been calculated as 3.4 lb. His incorrect value of 6.9 lb was reasonably close to the observed value of 6 lb, and he attempted to explain the discrepancy by assuming a time-dependency for his material; a load of 300 lb might have broken the tension specimen after a long time, compared with the 330 lb observed.

Mariotte had, as a result of his experiments on wood, a convincing refutation of Galileo's theory, but, because of his mistake in calculation, only fortuitous support for his own theory. If the fracture strength of a wooden beam in bending can be predicted from an elastic distribution of strain coupled with a maximum-strain postulate of failure, then it will be seen from Table 2.1 that the 'correct' load at which the beam should have broken is 2.6 lb. On the other hand, if failure is supposed to occur by the formation of a plastic hinge in accordance with modern simple plastic theory, then the relative strength for the circular specimen would be predicted as $2Sd/3\pi$, leading to a breaking load of the cantilever of 4.4 lb, still not very close to the observed load.

Mariotte also carried out a purely 'structural' test as against a 'strength of materials' test. He found that a circular glass rod over a 9 in span carried a central breaking load of 1 lb 10 oz 5 drams. The test was repeated on a similar rod whose ends had been carefully bound with twine before insertion into mortises; the breaking load of this 'fixed-ended' beam was 3 lb 5 oz 4 drams, almost exactly double that of the simply supported beam. He concluded that an encastré beam has twice the strength of the corresponding simply supported beam, a conclusion that will be approved both by the modern conventional elastic designer and by a designer versed in plastic methods.

2.2 Leibniz and James Bernoulli

Mariotte's book was published posthumously in 1686, but the work was known by 1680, and Mariotte's disagreement with Galileo stirred the interest of Leibniz – he set out to resolve the dispute in a paper published in 1684 in the *Acta Eruditorum* of Leipzig. Leibniz examined the 'inextensional' and 'extensional' theories of Galileo and Mariotte respectively, and concluded that Mariotte was correct; he placed the neutral axis at the bottom of the section, and deduced an 'elastic' distribution of stress. Leibniz added little that was new, although Girard considers that the 1684 paper provides good evidence that Leibniz could have made a deeper contribution had he so wished; but, as Leibniz said himself, he would leave it to others to uncover the basic theory. It seems clear that Leibniz had no real interest in the problem, probably because there was little mathematical content. His closing remarks indicate his belief in the power of mathematics to solve physical problems: *'ut proinde his paucis consideratio tota haec materia redacta sit ad puram Geometriam, quod in physicis et mechanicis unice desideratur'*. As Girard puts it, *'il préféra d'exercer son génie à des spéculations plus sublimes'*.

The linear assumption of the Mariotte–Leibniz theory (as it came to be called) was questioned by Varignon in 1702. He continued to place the neutral axis at the base of the section, but he developed what might be called a unified theory of bending. He saw no reason to accept *a priori* that stresses should be linearly dependent on strains (to continue to use anachronistic words), and proposed instead that fibre stress should be expressed as a general function of distance from the neutral axis. As a general expression he discussed (in effect) the formula $\sigma = k\epsilon^m$ as a representation of the stress–strain law; for $m = 0$, the stress is constant, and Galileo's formula is recovered, while for $m = 1$ the linear theory of Mariotte is obtained. Varignon made the general integration of his formula in order to obtain the value of the moment of resistance, but carried the work no further.

However, this contribution by Varignon provoked another giant into entering the discussion. James Bernoulli had been interested as early as 1691 in the problem of the shape of a bent elastic member, that is, in the study of the *stiffness* of beams, discussed more fully in Chapter 4. He was, perhaps, irritated (as a mathematician) at the inexactness of the attack on the question of the *strength* of beams, and, in 1705, in his *Véritable hypothèse de la résistance des solides*, he set out to deal with the matter. He sees no reason to adopt a linear theory, and he claims to be the first to consider compressions as well as extensions, although Mariotte had discussed this 20 years earlier. Indeed, James Bernoulli repeats Mariotte's precise mistake with the factor of 2, which leads him to conclude that the neutral axis can be placed anywhere; that is, the position is indifferent. Bernoulli's Mémoire of 1705 is a revised version, post-Varignon, of an earlier 1694 Leipzig paper. Like Leibniz, Bernoulli contributes little; all that he really shows is that the linear theory of Leibniz is no more admissible than the uniform theory of Galileo. Nevertheless, the fact that two such scholars should write on the problem of bending testifies both to its importance, and to an awareness that the solution was not easy.

2.3 Parent 1713

The academician Parent made a substantial contribution to the problem of bending (and to other branches of engineering), although his work was almost completely ignored at the time, and continued to be little noticed. He never advanced above the lowest grade of the *Académie*, that of *élève* (there were 20 each of *pensionnaires, associés* and *élèves*, together with 10 *honoraires*). Parent published a series of Mémoires in the first decade of the eighteenth century, and he collected and expanded these in three volumes of *Essais et recherches* in 1713. These volumes are very small (duodecimo) and very thick, and do not encourage close attention; this reason for neglect was advanced in the official obituary in the *Histoire de l'Académie* for 1716.

Parent's earlier papers on bending continue to place the neutral axis at the base of the section, but he reviewed the whole theory in the collected essays. A Mémoire in vol. 2 is entitled '*Comparaison des résistances des Cylindres & segmens pleins, avec celles des creux égaux en base, dans le systême de M. Mariotte*', that is, Parent is concerned with beams of both solid and hollow cross-section. He comments first on Mariotte's use of theory for a rectangular beam to explain results obtained from a beam of circular cross-section. He makes the integration over the circle (still with the neutral axis at the base) and obtains a value for the relative strength of $\frac{5}{16}Sd$ instead of the value $\frac{1}{3}Sd$ used by Mariotte. Parent observes, justly, that had Mariotte used the slightly smaller value, the theoretical breaking load (6.4 lb, cf. Table 2.1) would have been even closer to that observed (6 lb).

However, Parent had already discovered and corrected Mariotte's mistake of the factor 2 in the value of the relative strength calculated with respect to a central neutral axis; he had arrived at the value of $\frac{1}{6}Sd$ for a rectangular section and he notes that Mariotte should have calculated the breaking load as half the value shown in Table 2.1, that is, $3\frac{1}{2}$ lb, 'ce qui seroit bien éloigné de la verité'.

Later in this same Mémoire Parent discusses in physical terms the position of the neutral axis. Just before fracture the extreme tensile fibre will be the most strained, and the behaviour of that one fibre will influence the behaviour of the whole section. During the process of fracture, as successive fibres reach their limiting strains, the neutral axis will descend, until it reaches the base of the section, but it was clear to Parent that it will not be found there *before* fracture. Thus he distinguished clearly between the (elastic) working state of the beam, and the ultimate condition which is governed by the weak tensile behaviour of the material – and he is aware that the neutral axis can shift between the one state and the other.

Parent finally, and for the first time, uses rational mechanics to determine the position of the neutral axis, rather than placing it in a 'likely' position. The work is contained in the 14th Mémoire of vol. 3 of the *Essais*: '*De la véritable méchanique des résistances relatives des Solides ...*'. As every present-day first-year student of statics knows, three equations must be written to express the equilibrium of a two-dimensional rigid body. The equations can be written in different ways; in a form convenient for Galileo's beam, forces must sum to zero vertically, and also horizontally, and there must be no net moment acting on the beam. Galileo had achieved moment equilibrium by equating the couple exerted by the applied load acting about the fulcrum (the neutral axis at the base of the section) to the couple exerted by the fibres of the beam all at their fracture limits. He had apparently ignored the other two requirements, that forces must sum to zero both vertically and horizontally. The neglect may indeed be only apparent; Galileo would perhaps not have thought it necessary to remark that a vertical force must be generated at the root of his cantilever beam.

Parent writes the all-important equation expressing horizontal equilibrium; there is no horizontal force acting on the beam, so that there can be no net horizontal force acting at the root of the cantilever. He takes a general linear strain distribution, fig. 2.3, and allows unequal elastic moduli in tension and compression; the tensile stress in fibre AT will govern fracture, but the compressive stress in fibre BX may well be larger. Whatever their values, Parent states that the total *résistance* of the fibres of the compressive triangle CBX must equal that of the tensile triangle CAT, these two forces acting through I and D respectively, '*qui est une*

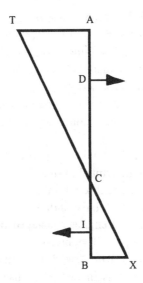

Fig. 2.3. Parent's distribution of bending strain.

proprieté dont personne n'avoit encore parlé'. He is fully aware that his correct statement of horizontal equilibrium is breaking new ground.

In using the distribution of fig. 2.3 to explain Mariotte's test result, Parent shows that the relative strength in bending is proportional to AC, that is, to the distance of the extreme fibre from the neutral axis of the beam. In one calculation, for example, he deduces that the ratio AC/AB in fig. 2.3 should be 9/11 to account for Mariotte's test, that is, the neutral axis at fracture is close to the bottom fibre. In fact, the calculations do not agree well with the experimental result, and it seems that, in a series of closely written essays, Parent was moving towards the conclusion that the calculation of the bending strength of a beam was not necessarily a simple matter. Parent was, so to speak, forced to consider linear-elastic behaviour but with unequal moduli in tension and compression; similarly, Bülffinger (1729) revived Varignon's idea that stress might be a non-linear function of strain. Mariotte's results of fracture of beams were becoming less, and not more, explicable in terms of a simple elastic theory of bending.

2.4 Bélidor 1729

Parent did not have a range of experimental results against which to match his theory, but in the next 40 years many tests on wood, stone and iron are reported. In 1729 Musschenbroek published his famous *Physicae experimentales* ..., giving the results of a large number of material tests in tension, compression and bending, mainly on a variety of types of wood; some tensile tests on metal wires are also reported. These tests were carefully made, and full descriptions and engravings of the testing machines are given. The tests are also notable for being apparently the first on slender columns, where failure occurs by buckling rather than fracture of the material.

Bélidor's enormously influential *La science des ingénieurs* was also published in 1729. Although printed in quarto, the volume is effectively a handbook, and its six parts, each containing between 60 and 100 pages, are numbered separately, and have individually engraved first pages. Bélidor is writing for the civil engineer engaged in practical design, and he deals with the major topics of eighteenth century construction. Thus book 1, *de la Théorie de la Maçonnerie*, deals with the design of retaining walls to hold back soil, and is in fact a treatise on the thrust of soil, that is, on soil mechanics. However, the theory is not original; Bélidor makes little scientific advance, but is concerned to establish the tables of thickness and batter of retaining walls with which he concludes the book.

Similarly, in book 2, *de la Mécanique des Voûtes*, Bélidor follows La Hire's analysis of 1712; the work is directed towards the calculation of the thrust of arches, so that abutments for bridges can be designed. (The development of arch theory proceeded almost independently of other structural analysis, and is described in Chapter 5 below.) In both these books Bélidor gives an early example of the engineer distilling the findings of scientific investigation into rules of design which form a 'code of practice'.

In book 3 Bélidor gives a qualitative discussion of the properties of building materials, including stone, brick, lime, sand, pozzolana, plaster and mortar. Unit weights are given for a range of materials, including metals (iron, brass, copper, lead), various sands and clays, brick, various building stones, and several kinds of wood, but no other quantitative properties are given; strengths are not mentioned.

Bélidor had been teaching for a few years at the artillery school at La Fère, and in book 4 he reports the results of tests on the breaking of beams made by some of the students (who were, of course, army officers). The tests were apparently designed to illustrate his lectures, and do not seem to have been well carried out; there were eight experiments each on three identical specimens. The tests on simply supported beams confirmed Galileo's finding that strength is proportional to the square of the depth; the tests on fixed-ended beams show only that Bélidor found it difficult to provide sufficiently stiff fixing to cause fracture at both ends and centre. Once again, Bélidor summarizes theory but makes no advances. He places the neutral axis in the surface of the beam. Book 5 deals with 'architectural' matters, such as the classical Orders, taper and entasis of columns, and so on. Book 6 outlines the preparation of specifications and contracts.

Bélidor's 'codification' of civil engineering knowledge did not prevent scientific advance elsewhere (his book was reprinted continually for a century, and his *Architecture hydraulique* was equally popular in the field of mechanical engineering). Other, better, material tests were made; Buffon (1740, 1741) investigated a large range of wooden beams (and seems to have been the first to note deflexions), and he also made some tests on iron rods. Poleni (1748) tested the strength of iron in his investigation of the reinforcement required for the dome of St Peter's, Rome; see Chapter 5 below. It was recognized that knowledge of the basic strength of the materials used was necessary to solve the problems of the strength of beams and the strength of columns.

2.5 Coulomb 1773

The strength of beams and the strength of columns are two of the four 'classic' eighteenth-century problems of civil engineering. The other two topics are the thrust of soil and the thrust of arches, and all four are addressed by Coulomb in his first published Mémoire, that of 1773. Coulomb had possibly been through the school at La Fère, some 40 years after Bélidor had started teaching there; he certainly joined the graduate school, the *École du Corps Royale du Génie*, in Mézières, in 1760. It is likely that Bélidor was the text-book at the school; Coulomb knew of the earlier work of Vauban (1704, 1706), who had tabulated thicknesses of retaining walls, and of the tests made by Musschenbroek. His knowledge of the science of civil engineering was, however, probably not much wider than these references imply.

Coulomb was posted to Brest after graduation. In February 1764 a ship was sailing for Martinique, and the engineering officer who had been assigned to overseas duty fell ill. Coulomb was drafted in his place at very short notice, and it was not until June 1772 that he managed to return to France. His duties on Martinique included the design of fortifications to defend the island against possible renewed attack by the English, and it is perhaps not surprising that, as a young graduate from university, he found that what he had been taught was insufficient for his tasks. It was during the eight years on Martinique that Coulomb tried to find solutions to the four classic problems; he wrote the Mémoire, dated and presented in 1773 but published in 1776, for his own use (as he says himself in the introduction). He hopes the *Académie* will find his small contribution to the monument of learning to be useful; the grand design is in the hands of great men, but lesser workers, hidden in the darkness of the foundations, may also perhaps be of help.

The most famous section of the Mémoire is that which deals with the thrust of soil, and Coulomb is regarded as the founder of the science of soil mechanics – in fact, rational mechanics had been applied to the problem since at least 1691 (Bullet). However, Coulomb's fundamental contribution was to assume that failure occurred by shear along a plane, and to determine the position of that plane by the use of 'principles of maximum and minimum'.

In dealing with the problem of bending, Coulomb was influenced by his teacher at Mézières, Charles Bossut. With the hindsight of the 'correct' elastic solution, in which the neutral axis is central for a symmetrical cross-section, it appears that Coulomb was trying to solve the elastic problem. The short sections in the Mémoire, however, state clearly that they deal with the *fracture* of bodies. Coulomb presents two theories, one for wood and one for stone, and he was almost certainly ignorant of the work of Parent 60 years earlier. Instead, following Bossut, he assumes that, at fracture of a wooden beam, stresses are proportional to distance from the neutral axis; almost by accident he obtains what is now thought of as the elastic solution to the problem. For stone he makes the alternative (Galilean) assumption that all the 'fibres' fracture in tension at the same stress, and he moves the neutral axis to the surface of the beam.

However, Coulomb starts his bending analysis by considering a general stress distribution, fig. 2.4 (Coulomb's Fig. 6). He makes a hypothetical vertical cut ACD of the beam (the letter D is missing from the figure), and he states very clearly the equilibrium conditions that must be satisfied at this cross-section. The portion CA above the neutral axis will be acted upon by tensile forces in the direction QP, while the portion CD is acted upon by compressive forces Q'P'. The forces QP are resolved into components QM and MP, where QM (in modern terminology) represents a shear stress and MP a longitudinal stress. Thus the portion ADKL of the beam is acted upon by all the horizontal forces MP, by all the vertical forces QM, and by the weight ϕ. Since this portion of the beam

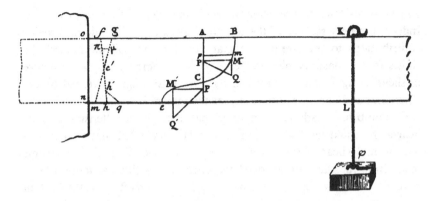

Fig. 2.4. Coulomb's Fig. 6: equilibrium of a loaded cantilever.

is in equilibrium, the sum of the horizontal forces is zero, so that the total tensile force corresponding to the block CAB must equal the total compressive force CeD. Further, the sum of the vertical forces QM must equal the weight ϕ; and finally, by taking moments about C, Coulomb obtains the equation $\int$ Pp·MP·CP $= \phi$ LD.

In applying these equations to determine the fracture strength of a beam, Coulomb states clearly that the (shear) stresses QM have very little influence, provided that the lever arm nL of the weight ϕ is much larger than the depth on. The question of shear force is considered in more detail in Chapter 3 below.

Coulomb considers first 'a piece of perfectly elastic wood'; he assumes that the material extends in proportion to the load, and that tensile and compressive moduli are the same. Then the line fh close to the root of the cantilever (fig. 2.4) will move to gm under the action of the load; the stresses are represented by $\pi\mu$, and the tensile triangle fge must equal the compressive triangle hme. Coulomb immediately derives the 'correct' value of the elastic modulus, $\frac{1}{6}bd^2$.

Similarly, for stone, Coulomb assumes that the member is 'composed of stiff fibres, that can be neither compressed nor extended'; the body fractures by rotation about h, and the resulting section modulus is therefore determined as $\frac{1}{2}bd^2$. Thus Coulomb is considering the fracture of a body whose material has a certain strength in tension, and an infinite strength in compression. Coulomb is well aware that an infinite compressive strength is impossible, and that a finite area, hh' in fig. 2.4, will sustain the compressive load on the cross-section. This compressive load will, says Coulomb, cause fracture to occur along the plane h'q, and he deals later in the Mémoire with the way of finding this plane. Indeed, the next section of the Mémoire gives one of Coulomb's major contributions to the strength of materials (and to soil mechanics); he shows that a masonry pier, loaded axially in compression, will fracture in shear along a plane inclined to the axis at an angle governed by the (frictional and cohesive) properties of the material.

Coulomb had made tests on stone specimens, both in tension and (as near as he could manage) in pure shear. He concluded that strengths in tension and shear were very nearly equal. His bending tests on beams made from the same stone correlated fairly well with these tension tests, particularly when the neutral axis was allowed to shift from the bottom fibre h to a higher position h′. This is the point where Coulomb leaves the problem, to move on to questions of columns, soil and arches. He concludes that the breaking strength of a stone cantilever is governed by a moment of resistance $\frac{1}{2}Sd$ (Galileo's formula), and of a wooden cantilever (Coulomb does not appear to have made tests on simply supported timber beams) by $\frac{1}{6}Sd$ (Parent's formula). Coulomb offers no hint that an elastic stress distribution might be of interest in itself; indeed, the idea of an elastic working state of a beam was almost certainly not considered by him.

Coulomb read his Mémoire to the *Académie* on 10 March and 2 April 1773, no doubt on the recommendation of his old teacher Bossut, who was an *associé* of the *Académie* in the class of *géométrie*. It was the first of 32 memoirs to be presented by Coulomb between 1773 and his death in 1806, first to the *Académie* before it was abolished in 1793, and later to the reconstituted *Institut*. From his return to France from Martinique in 1772 Coulomb continued his career as an army officer in the *Corps Royal du Génie*, but Bossut reported (with Borda) very favourably on the first memoir, and as a result the memoir was published in 1776 and Coulomb was admitted to the *Académie* in 1774 as Bossut's *correspondant*.

Coulomb's careers in academe and in the army were not incompatible, but he moved slowly to devoting his full time to science; in 1781 he secured both full membership of the *Académie* and permanent posting to Paris, with the rank first of Captain and then of Major. He resigned from the army in 1790, and became a full-time salaried member of the *Académie* and later of the *Institut*.

2.6 The early nineteenth century

Coulomb never again took up the problem of bending (the famous memoirs on electricity and magnetism, for example, were written between

1785 and 1791) – nor, it seems, did anyone else in France during the last quarter of the eighteenth century. Girard's book of 1798 has been mentioned; he refers fulsomely to the work of Coulomb, but, like Coulomb, he had not discovered Parent's *Essais* of 1713, citing only some tests made by Parent in 1707 and 1708. Girard, in fact, makes no advance on Coulomb's study of 25 years earlier; he continues to place the neutral axis at the bottom face of the cross-section (fig. 2.2), and he could find no place for an 'elastic' theory, based upon a central position of the neutral axis, as an explanation of the experimental results. An elastic modulus of $\frac{1}{6}bd^2$ was useless as a predictor of fracture; the value of $\frac{1}{2}bd^2$ (Galileo) seemed best for stone and $\frac{1}{3}bd^2$ (Mariotte) for wood.

Citizen Girard's book had been read by citizen Coulomb before publication in Year VI (i.e. 1798); Coulomb and Prony had been charged by the *Institut* to report on the book. (Prony was one of the founders of the *École Polytechnique* in 1794, and he became Director of the *École des Ponts et Chaussées* in 1798.) Their view, that it was the most complete work on strength of materials from the viewpoint of both theory and experiment, is printed as a preface to the book. Coulomb signed this statement; he clearly supported the view that calculations for the fracture of beams should be based on the neutral axis lying in the surface of the cross-section.

Thus, enshrined in the approved theory of the *Polytechnique* and the *Ponts et Chaussées*, was a theory of bending, derived by rational mechanics by Coulomb, but represented by quasi-empirical formulae for the calculation of strength. Just as Coulomb had not known of the work of Parent, so Navier, who had graduated from the *Polytechnique* and who finally taught at the *Ponts et Chaussées*, was not fully aware of the work of Coulomb, nor that the position of the neutral axis was to be found by using the condition of no net thrust at a cross-section. Indeed, as late as 1819 Navier taught in his courses that the neutral axis was to be located by equating the *moment* of the compressive forces acting at a cross-section to the *moment* of the tensile forces; in this, he was apparently following a misreading of Coulomb by Duleau (1820). (It was not until the publication in 1826 of Navier's *Leçons* that (Coulomb's) correct restatement of horizontal *force* equilibrium was given, so that the position of the neutral axis in elastic bending could be located as passing through the centre of gravity of the cross-section.)

Thus the neutral axis had finally moved from the face of the cross-section, the position assumed by Galileo, Mariotte, (Coulomb) and Girard, to a 'central' position, that is, to the accidentally correct position for elastic bending of a symmetrical cross-section. French work was known in England, and in 1817 P. Barlow followed Duleau and pre-1826 Navier in equating the moments of the tensile and compressive stress blocks instead of the resultant forces; Barlow corrected this mistake in the 1837 edition of his book. However, the movement of the neutral axis was still not at this time a consequence of the search for an elastic solution to the problem of bending; it was the fracture problem that continued to be investigated. Thus Barlow had made tests on wooden beams (typically 2 inches square in cross-section and 48 inches long), and had observed that the fractures involved about $\frac{3}{8}$ of the section failing in tension and about $\frac{5}{8}$ in compression. By contrast, Tredgold (1822) had his own method of calculating the position of the neutral axis; the section moduli of the two halves of the section should have the same value, that is, the second moment of area of the tension side of the cross-section, divided by the maximum ordinate from the neutral axis, should be set equal to the similar expression for the compressive side. (As usual, this calculation leads to the 'correct' central position for symmetrical sections.)

Eaton Hodgkinson published two memoirs in 1824 and 1831 which make a substantial advance in the analysis of the problem. He knows of the work of Coulomb, and he constructs a general stress distribution to satisfy all of the equilibrium conditions, including that of no longitudinal force on the cross-section. He assumes that the tensile stress might be represented by the formula

$$\sigma = \sigma_0 \left(\frac{y}{a}\right)^n , \qquad (2.1)$$

where the symbols are shown in fig. 2.5; different constants (σ_0', a', n') hold for the material in compression. For a rectangular section of depth $d = a + a'$, Hodgkinson shows that horizontal equilibrium is satisfied if

$$\frac{\sigma_0 a}{n+1} = \frac{\sigma_0' a'}{n'+1} , \qquad (2.2)$$

and he obtains a general expression for the moment of resistance.

The constants are to be found experimentally and, once they are known, bending strengths can be predicted. As an example, Hodgkinson found that for Quebec oak, $n = 0.97$ (tension) and $n' = 0.895$ (compression); these results came from simple tensile and compressive tests. A bending test to fracture determined the position of the neutral axis from the experimental result $a/a' = 23/25$. The value of σ_0 was taken as 8000 lb/in^2; tensile strengths of some other materials are taken from Musschenbroek. Hodgkinson's bending formula was therefore essentially

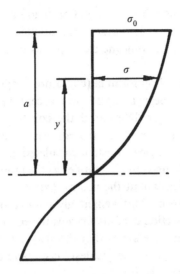

Fig. 2.5. Hodgkinson's distribution of bending stress.

empirical, although it incorporated fully the correct statements of equilibrium. It did not use 'compatibility' or 'stress–strain' relations, although as a matter of fact Hodgkinson assumed that strain in bending was proportional to distance from the neutral axis; thus the assumed power law was actually an assumed stress–strain relationship of the form $\sigma = k\epsilon^n$.

2.7 Navier 1826 and Saint-Venant 1864

As was noted above, Navier correctly located the position of the neutral axis for elastic bending in his 1826 *Leçons*. The analysis starts simply by statements of the three conditions of horizontal, vertical and moment equilibrium, and it follows that the neutral axis passes through the centre of gravity of the cross-section. Saint-Venant edited an edition of Navier in 1864, in which the text is expanded enormously by footnotes. As an example of these notes, Saint-Venant remarks that Navier's statement of vertical equilibrium at a cross-section of a beam is muddled about the way the shear forces act (the question of shear is discussed below in Chapter 3). Saint-Venant goes on to tabulate the assumptions that Navier has made, and notes that he is really dealing with pure bending in which shear is absent. Further, in his elastic analysis in which Navier derives the basic bending formula $M/I = E/R$, he has assumed tacitly that sections originally plane remain plane after bending, and that individual fibres are free to expand and contract without affecting their neighbours. Saint-Venant is, in fact, stating the assumptions of the simple theory of elastic bending.

The linear-elastic bending formula (curvature proportional to bending moment) had actually been assumed over a century earlier, as will be seen in Chapter 4. Navier determined the constant of proportionality (EI) in terms of Young's modulus (i.e. an elastic constant E from the tension test) and a geometrical constant calculated as an integral over the cross-section, which Saint-Venant identifies as the moment of inertia I (or second moment of area) about the neutral axis of bending. Navier had, then, with some help from Saint-Venant, given an account of the problem of elastic bending in section 3 of his *Leçons*; in section 4 he moves on to the problem of fracture. He states that the simplest hypothesis, and the one closest to reality, is that the greatest elastic strain, in either tension or compression, governs the fracture.

Navier remarks explicitly that his theory of fracture is based on the assumption that behaviour is linear-elastic right up to the point of failure. This is a clear statement that it is necessary to make only elastic calculations in order to determine strength. Those calculations will give a largest elastic strain at some point in the structure, and it is there that failure will occur. It is the linear-elastic analysis that gives the essential result for the computation of the moment of resistance of a beam: the section modulus can be written $z = I/a$, where a is the distance of the extreme fibre from the neutral axis, so that $M = \sigma_0 z$. Navier allows that a material might behave differently in tension and in compression, and this will involve a shift in the position of the neutral axis. (Saint-Venant digresses at this point to examine the flexure of sections made from material with such unequal elastic moduli.)

Thus the question of strength is an elastic problem, and Navier derives section moduli for the rectangle ($\frac{1}{6}bd^2$) and for the circle ($\frac{1}{4}\pi r^3$), and also for a rectangle bent about an inclined axis. This last analysis, as noted by Saint-Venant, is wrong; Navier assumes that the neutral plane of bending is always horizontal for vertical loading, whatever the orientation of the rectangular cross-section. Saint-Venant gives the correct analysis, and he acknowledges his debt to the lithographed notes of Persy, recording his lectures in 1834 for the *École d'Artillerie et du Génie* at Metz (Todhunter also consulted these lecture notes). Persy was apparently the first to add a *fourth* equation of equilibrium to the three demanded by the statics of a body considered in a plane – namely the equation resulting from moments taken about an axis at right angles to the axis of bending. The moment about this axis of the forces acting on the cross-section must also be zero, and Saint-Venant shows how this leads to the determination of principal axes, and that a correct general analysis of bending requires the loading to be first resolved into the principal planes.

As has been noted, section 4 of Navier's *Leçons* purports to deal with bending failure, but is largely a continuation of section 3 on elastic flexure. Saint-Venant does not explicitly disagree with Navier's view that bending failure may be examined by a linear-elastic theory, but he has a twelve-page footnote at this point in which the consequences of a non-linear theory are examined. Saint-Venant mentions the work of Varignon and Hodgkinson, and he proposes a formula similar to that of Hodgkinson, but in the slightly different form

$$\sigma = \sigma_0 \left[1 - \left(1 - \frac{y}{a} \right)^n \right] \, , \qquad (2.3)$$

where the symbols are, as before, defined in fig. 2.5, and different constants (σ_0' etc) are used for compression rather than tension. The rectangular section is examined, and longitudinal equilibrium leads to the condition

$$\frac{n}{n+1} a\sigma_0 = \frac{n'}{n'+1} a'\sigma_0' \; ; \qquad (2.4)$$

the value of the bending moment at the cross-section may be written

$$M = \frac{ba^2}{2}\sigma_0 \frac{n(n+3)}{(n+1)(n+2)} + \frac{ba'^2}{2}\sigma_0' \frac{n'(n'+3)}{(n'+1)(n'+2)} \, . \qquad (2.5)$$

Saint-Venant discusses several cases. First, he takes $n = n' = 1$, so that $a\sigma_0 = a'\sigma_0'$, and the stress distribution is as sketched in fig. 2.6. Since $(a + a') = d$, the value of the bending moment may be written

$$M = \left(\frac{a}{d} \right) \left(\frac{1}{3}bd^2\sigma_0 \right) \, . \qquad (2.6)$$

Thus for the neutral axis on the centre line, $a/d = \frac{1}{2}$, and the elastic modulus of $\frac{1}{6}bd^2$ is recovered. For the neutral axis dropping to the bottom of the section, $M = \frac{1}{3}bd^2\sigma_0$, and Mariotte's formula results, in which it is assumed that tensile stress governs the fracture, the material being able to resist very large compressive stresses.

For the more general group of non-linear cases, Saint-Venant assumes that the stress diagram in bending has continuity in slope at the neutral axis, that is,

$$\frac{n\sigma_0}{a} = \frac{n'\sigma_0'}{a'} \, . \qquad (2.7)$$

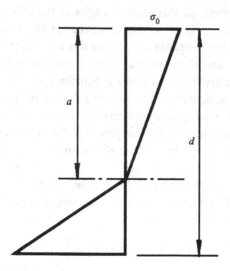

Fig. 2.6.   Linear distribution of stress with unequal tensile and compressive moduli.

The value of the bending moment may then be calculated as

$$M = \frac{1}{2}bd^2\sigma_0 \frac{1}{(1+k)^2} \left(\frac{n}{n'+1}\right) \left[\left(\frac{n+3}{n+2}\right) + \frac{1}{k}\left(\frac{n'+3}{n'+2}\right)\right] , \qquad (2.8)$$

where

$$k = \frac{a}{a'} = \left(\frac{n+1}{n'+1}\right)^{\frac{1}{2}} . \qquad (2.9)$$

For the symmetrical distribution, fig. 2.7, $n = n'$, $k = 1$ and $a = a' = \frac{1}{2}d$. The value of the fracture moment is

$$M = \frac{1}{4}bd^2\sigma_0 \left(\frac{n}{n+1}\right)\left(\frac{n+3}{n+2}\right) . \qquad (2.10)$$

The value $n = 1$ gives the linear-elastic case $M = \frac{1}{6}bd^2\sigma_0$; as the value of n increases, so the value of M approaches $\frac{1}{4}bd^2\sigma_0$, which is the maximum moment of resistance of a beam of rectangular cross-section made from perfectly plastic material. Figure 2.7 is sketched for $n = 3$; Saint-Venant sketches similar curves for various values of n, including the value $n = 10$, for which the value of M is less than 2 per cent below the full plastic value. Although Saint-Venant did not refer explicitly to the idea of perfect plasticity, he had nevertheless derived the plastic section modulus.

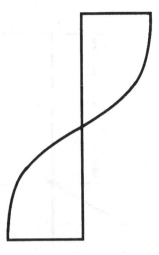

Fig. 2.7. Non-linear distribution of stress, equation (2.3), for the case $n = n'$, $a = a'$ (the sketch shows the distribution for $n = 3$).

Finally, Saint-Venant considers the case for which $n' = 1$, n is allowed to vary, and the elastic moduli are the same in tension and compression for very small strains (fig. 2.8). The various quantities become

$$\left.\begin{array}{l} k = \left(\dfrac{n+1}{2}\right)^{\frac{1}{2}}, \quad a = \left(\dfrac{k}{1+k}\right)d, \\[3mm] M = \dfrac{1}{4}bd^2\sigma_0\dfrac{1}{(1+k)^2}n\left[\left(\dfrac{n+3}{n+2}\right) + \dfrac{4}{3k}\right]. \end{array}\right\} \tag{2.11}$$

The explicit assumption is that fracture is governed by the maximum tensile stress, and that compressive strength is not of significance. As before, the condition $n = 1$ gives the elastic case $M = \frac{1}{6}bd^2\sigma_0$. As n increases, the value of a approaches the full depth d and, in the limit, Saint-Venant obtains Galileo's formula $M = \frac{1}{2}bd^2\sigma_0$.

Saint-Venant points out that for $n, n' > 1$, the section modulus always lies between the limits $\frac{1}{6}bd^2$ (Coulomb) and $\frac{1}{2}bd^2$. He proposes that the formula $\frac{1}{6}\alpha bd^2\sigma_0$, where α lies between 1 and 3, will serve as an empirical expression for predicting rupture moments. As an example, the value of α for cast iron is about 2, so that the value of n for the stress distribution of fig. 2.8 would be between 5 and 6; the theory could then be applied to the bending of non-rectangular sections, using these experimentally derived values of the constants.

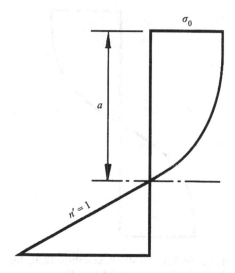

Fig. 2.8. Non-linear distribution of stress for a material strong in compression.

Thus, whereas Navier in 1826 was firmly expounding the linear-elastic theory of bending as also serving to explain fracture, Saint-Venant, in his 1864 notes, explored various semi-empirical non-linear theories as being more truly applicable to the experimental results. It is not surprising that the resulting formulae were, in a practical sense, successful. The general stress distribution of fig. 2.5 must be such that there is no net thrust across the cross-section, so that one condition must hold between the six empirical constants $(\sigma_0, a, n; \sigma_0', a', n')$. If the stress distribution is related to bending strains (plane sections remain plane) then a further condition may be established by assuming that elastic behaviour is identical in tension and compression (no discontinuity in slope at the neutral axis in fig. 2.5). This still leaves four unknown empirical constants to be determined from the experimental results, and an enormous range of data can be 'fitted' with such freedom in the parameters.

2.8 The full plastic moment

It was, of course, Navier's linear-elastic philosophy that became paramount and was virtually unquestioned for a century and a half as the correct approach to structural design. Non-linear ideas, however, were not lost sight of, although if they were mentioned, they were treated mainly as a matter of scientific curiosity. Ewing's undergraduate text of

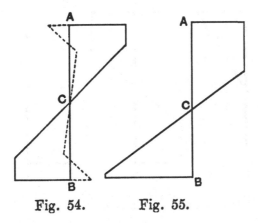

Fig. 54.   Fig. 55.

Fig. 2.9. Ewing's stress distributions for bending beyond the elastic limit.

1899, for example, gives an almost exclusively elastic account of *The strength of materials*. He does indeed discuss the ultimate strength and non-linear strain of tension specimens, but the structural applications are all linear. The single exception may be found in a page mentioning the 'influence of bending beyond the elastic limit on the distribution of stress', and fig. 2.9 reproduces two of Ewing's illustrations concerned with the bending of a beam of rectangular cross-section.

In his Fig. 54 Ewing imagines the material to be 'strictly elastic up to a certain limit of stress, and then so plastic that any small addition to the stress produces a relatively very large amount of strain'; the state of stress of the partially plastic section is shown, together with the state of internal (self-equilibrating) stress that would remain when the beam is relieved from external load. In the other sketch shown in fig. 2.9 (Fig. 55) the material is supposed to have different elastic limits in tension and compression, resulting in a shift of the neutral axis after first yield. Ewing gives no numerical work to correspond to the distributions of fig. 2.9, although it seems clear that he had made the appropriate calculations.

The diagrams of fig. 2.9 are, of course, forerunners of the now familiar 'elastic/perfectly plastic' idealization of material behaviour for mild steel; as the bending moment on a cross-section is increased, fig. 2.10, so the material passes from an elastic state through a partially plastic state to a condition of full plasticity, where the maximum bending moment, the 'full plastic moment', is attained. Robertson and Cook published in 1913 the results of a comprehensive series of tests on mild-steel specimens, and they

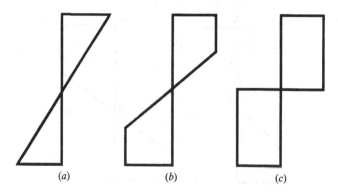

Fig. 2.10. Elastic/plastic bending; (*a*) elastic limit, (*b*) partially plastic, (*c*) fully plastic.

propounded a simple theory of plastic bending based on their work. As a consequence it was now possible to calculate theoretically the behaviour of Galileo's cantilever beam; a complete load/deflexion curve could be predicted as the tip load increased slowly until collapse occurred.

The essential concept for structural design, however, is not a particular loading curve, but the fact that there is a limiting value of the moment of resistance, the full plastic moment. The idea is particularly simple for a material like mild steel but, as Saint-Venant saw, an empirical theory could be adapted to predict the ultimate behaviour in bending of a beam made of any other material. The ways in which plastic theory came to influence the theory of structural design are discussed in Chapter 9.

2.9 Axial load

Knowledge of the value of the full plastic moment, then, is needed for the simple plastic theory of structural design. It was realized early in the development of that theory that the presence of an axial load could reduce the value of the full plastic moment. Analysis of the effect is straightforward if the usual assumptions of the engineering theory of bending are made. (A major consequence of the presence of a compressive axial load is, of course, that the member may buckle and move into a pattern of behaviour not so far considered; the question of the elastica forms part of the subject matter of Chapter 4. Discussion here is confined to *local* effects, that is, to the calculation of the value of the moment of resistance.)

Galileo's beam is shown in fig. 2.11, but the tip is now acted upon by an axial load *P* as well as the transverse load *S*; it is assumed that

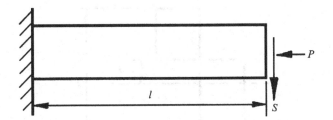

Fig. 2.11. A cantilever beam acted upon by a transverse and an axial load.

the beam does not buckle. Values of P and S are sought that will cause collapse of the beam, that is, loads that will cause a plastic hinge to form at the root of the cantilever. If the transverse load S is replaced by the variable $M = S\ell$, then the 'yield surface' that connects the values of M and P is to be determined. (In this simple analysis it is assumed tacitly, following Navier, that the presence of the shear force S itself acting at the root of the cantilever has no effect on the value of the full plastic moment. This approximation may be valid for long beams, that is, for relatively low values of S, but the problem is complex, and is discussed separately in Chapter 3 below.)

In the absence of axial load, the full plastic stress distribution in bending is as sketched in fig. 2.10(c), leading to a full plastic moment of value M_0. The presence of an axial load P distorts the bending-stress diagram to that sketched in fig. 2.12. For a rectangular section, the value of M is given by $M = \left(1 - \alpha^2\right) M_0$, or, using a non-dimensional expression $m = M/M_0$ for the bending moment,

$$m = 1 - \alpha^2 \ . \tag{2.12}$$

Similarly the value of $p = P/P_0$ is given by

$$p = \alpha \ , \tag{2.13}$$

where P_0 is the value of the 'squash load', that is, the maximum axial load that can be imposed on the cross-section in the absence of bending moment. Thus the required equation of the yield surface, for m positive, is

$$m + p^2 = 1 \ , \tag{2.14}$$

and this curve is sketched in fig. 2.13, together with the corresponding (reflected) curve for m negative.

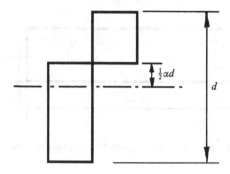

Fig. 2.12. Stress distribution for full plastic moment in the presence of axial load.

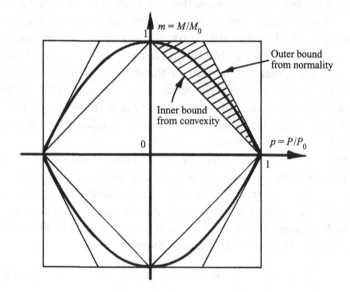

Fig. 2.13. Yield surface for rectangular section subject to bending and axial load.

As will be seen in Chapter 9, Gvozdev (1936) characterized the properties of yield surfaces such as that of fig. 2.13. Firstly, they must be
convex; that is, for this symmetrical problem, the equation connecting m
and p (i.e. equation (2.14)) must be within the square $m = \pm 1$, $p = \pm 1$,
and outside the diagonal square marked 'inner bound'. In fact the outer
bound can be narrowed by virtue of the 'normality rule' of plasticity, also
to be discussed in Chapter 9; whatever the cross-section of the beam,
it may be shown by consideration of simple deformation patterns that
the yield surface must be horizontal for $p = 0$ and must make an angle

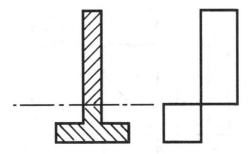

Fig. 2.14. The plastic zero-stress axis is an equal-area axis.

of $\tan^{-1} 2$ (just over 60°) with the p-axis for $m = 0$. (The yield surface for an I-section lies between the heavy curve shown for the rectangular section and the outer bound.) Secondly, the fact that the yield surface shown in fig. 2.13 is of quadratic form and is horizontal at $p = 0$ implies that moderate values of axial load will have only a small effect on the value of full plastic moment (and proportionately less for a practical I-section than for a rectangular section); this is of obvious help in the straightforward plastic design of engineering structures.

The ideas presented in this discussion on the influence of axial load, but without the notion of convexity of the yield surface, were developed by the 'Cambridge' team working in the 1940s and 1950s; the early work is summarized conveniently in Baker *et al.* (1956). Further discussion of the Cambridge work will be found in the next chapter and in Chapter 9.

2.10 Plastic bending about two axes

It has been seen that Navier had not discussed the idea of principal axes for elastic bending; it was Saint-Venant, following Persy, who formulated and solved the problem correctly. There is a corresponding difficulty for plastic bending – that is, in describing the formation of a full plastic moment when a cross-section is bent about an arbitrarily inclined axis. Brown (1967) first recorded the general features of this problem.

For a cross-section having at least one axis of symmetry, the 'principal' plastic axes will be orthogonal, and one will lie in the plane of symmetry. They are located by the requirement that, for no net thrust over the section, one half must be yielding in tension and one half in compression. In fig. 2.14, for example, the zero-stress axis is an 'equal-area axis', the two shaded areas of the T-section being equal. Thus the plastic principal

The Moment of Resistance

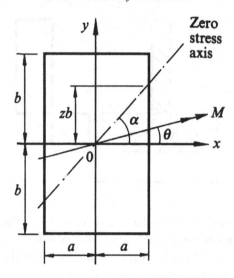

Fig. 2.15. A rectangular cross-section bent about an inclined axis.

axis in this case does not coincide with the elastic principal axis (which must pass through the centre of gravity of the cross-section).

If the bending moment is applied about an axis which is not one of the principal plastic axes, then, just as for the elastic problem, the zero-stress axis will not in general coincide with the bending axis. In fig. 2.15 for the rectangular section, the moment M acts about an axis inclined at θ to the co-ordinate axis Ox, while the zero-stress axis, which must of course pass through the centre of the cross-section (in order to give equal areas in tension and compression), makes an angle α with Ox. The section may be thought of as being acted upon by two independent moments M_x and M_y, where

$$\left.\begin{array}{l} M_x = M \cos \theta = 2ab^2\sigma_0 \left(1 - \tfrac{1}{3}z^2\right), \\ M_y = M \sin \theta = 2a^2b\sigma_0 \left(\tfrac{2}{3}z\right), \end{array}\right\} \tag{2.15}$$

where the notation is given in fig. 2.15 and σ_0 is the value of the yield stress of the material. Equations (2.15) are the parametric form of the yield surface

$$\left(\frac{M_x}{2ab^2\sigma_0}\right) + \frac{3}{4}\left(\frac{M_y}{2a^2b\sigma_0}\right)^2 = 1. \tag{2.16}$$

From fig. 2.15,

$$\tan \alpha = \frac{b}{a}z, \tag{2.17}$$

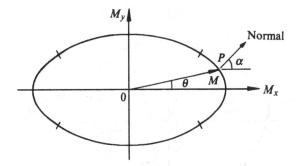

Fig. 2.16. Yield surface for the rectangular cross-section.

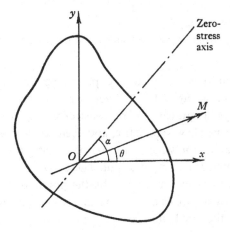

Fig. 2.17. General cross-section.

and equations (2.15) and (2.16) hold for $-1 \leq z \leq 1$, that is, for $|\tan \alpha|$ less than b/a. Similar equations may be derived by the interchange of a and b when $|\tan \alpha|$ is greater than b/a. The yield surface corresponding to equation (2.16), together with the similar equation for $|\tan \alpha|$ greater than b/a, and both of these with signs reversed (bending in the opposite sense), are plotted in fig. 2.16. The values of θ and α marked in fig. 2.16 are the same as those of fig. 2.15; Gvozdev's normality rule, mentioned above and to be discussed in Chapter 9, identifies the direction of the zero-stress axis. It will be seen that the values of θ and α are the same for $\theta = 0$ and $\pi/2$; the principal axes for plastic bending of the rectangular section are, as to be expected, the axes of symmetry.

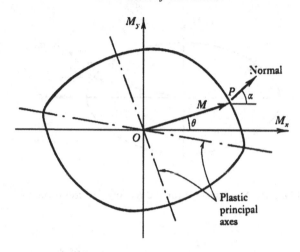

Fig. 2.18. Skew-symmetrical yield surface for the general cross-section.

Brown extended these ideas to the plastic bending of the general unsymmetrical section, fig. 2.17. The zero-stress axis is again an equal area axis, and a yield surface may be constructed as shown in fig. 2.18. This surface must be skew symmetric, and there are two values of θ for which $\alpha = \theta$, that is, for which the axis of deformation is parallel to the axis of the applied bending moment. These plastic principal axes are sketched in fig. 2.18; they are located by the points of tangency of the inscribed and escribed circles centred on the origin. The axes are clearly not necessarily orthogonal; nor need there be any coincidence with an elastic principal axis.

3

The Effect of Shear

The problem of the breaking strength of a beam continued to be visualized in the form stated by Galileo, namely that of a cantilever beam encastred at its left-hand end and loaded by a single weight at the free end. From this formulation was abstracted the 'cleaner' problem of the calculation of the breaking resistance of the cross-section adjacent to the support, since clearly this was the critical section of the beam.

In calculating the moment of resistance of the beam, Galileo considered only one of the three statical equations (or four, since Persy's contribution of 1834 must be included), namely that the moment of the forces acting at the cross-section must equal the moment of the applied load. He did not write the equation of longitudinal equilibrium (Parent (1713)) which helps to determine the location of the neutral axis of bending, nor did he resolve forces vertically, which leads to the idea of a shearing action on the critical section.

As has been seen, Coulomb (1773) did realise that the forces acting on the critical section must have vertical components in order to balance the load applied to the tip of the cantilever. Indeed two of Coulomb's four problems (the strength of columns, the thrust of soil) are concerned with shear fractures, and he tried to test his (stone) cantilever beam in pure shear by applying the load as close as he could to the encastred end. The experimental technique was not good, but Coulomb measured to his own reasonable satisfaction the strength of stone in pure tension and in pure shear, and related these two strengths by 'Coulomb's equation', involving two physical parameters, cohesion and friction.

3.1 Navier 1826, 1833

For Galileo's problem, Coulomb stated that the shear stresses at the critical section would have very little influence on the strength of the

43

beam, provided that its length were much larger than its depth. Navier
repeated this proviso, and originally (1826) had nothing more to say
on the matter. However, as Saint-Venant points out in his extended
notes to the 1864 edition, Navier added short paragraphs (§§152–155,
see Table 3.1 below) to the 1833 revision of the *Leçons*. These were in
response to the criticism that his theory did not help with the analysis
of such structural elements as lugs, hooks, gudgeons, keys and cotters,
to which list we may well add (says Saint-Venant) brackets, flanges,
studs, pins of pulleys and blocks, gear teeth, collars, screw threads and
rivets.

It would seem that Navier was not really interested in these 'mechanical-
engineering' structural applications, and his added paragraphs of 1833
continue to deal only with Galileo's problem. At the encastred end
of the beam the vertical load P will produce a 'vertical strain'
(*allongement*), and clearly (*il est naturel d'admettre*) the magnitude of
the load P will be proportional to the magnitude of this 'strain' and to
the cross-sectional area of the beam. An elastic constant is introduced
to represent this proportionality and, in accordance with his assumption of a greatest elastic strain, Navier states that fracture will occur
when the 'vertical strain' reaches a limiting value. In a notation not
used by Navier, there is therefore a maximum shear load P_0 that can
be imposed on the beam, and this value of P_0 is proportional to the
greatest 'vertical strain' that can be resisted by the material of the cross-
section.

For a long beam, the elastic bending moment M at the root of the
cantilever will give rise to a horizontal strain, and fracture will occur
when that (elastic) horizontal strain reaches a limiting value. Thus (again
in a notation not used by Navier), there is a maximum bending moment
M_0 that can be imposed on the cross-section of the beam in the absence
of shear force.

For the general case Navier has to deal with a horizontal strain and
a 'vertical strain' acting simultaneously at the encastred root of the
cantilever. Navier treats these strains as vectors; that is, the maximum
strain has value equal to the square root of the sums of the squares
of the horizontal and vertical strains. From this Navier deduces the
circular criterion of failure of the beam under the simultaneous action
of a bending moment M and a shear force P:

$$\left(\frac{P}{P_0}\right)^2 + \left(\frac{M}{M_0}\right)^2 = 1, \tag{3.1}$$

where, if ℓ is the length of the cantilever,

$$M = P\ell, \tag{3.2}$$

so that

$$P = \frac{P_0 \dfrac{M_0}{\ell}}{\left(P_0^2 + \dfrac{M_0^2}{\ell^2}\right)^{\frac{1}{2}}}. \tag{3.3}$$

As Navier points out, for large values of ℓ equation (3.3) gives $P\ell = M_0$, while for ℓ very small, $P = P_0$.

3.2 Saint-Venant 1855, 1856, 1864

Navier has not, of course, distinguished longitudinal strain from shear strain, and is not aware that strain cannot be treated as a vector. Saint-Venant's notes to the 1864 edition of the *Leçons* correct all this.

Saint-Venant's two great memoirs, on torsion and on flexure, were published in 1855 and 1856. The first deals with what has come to be known as 'Saint-Venant's problem', namely the distribution of stress in a prismatic member of defined cross-section under the action of torsional couples. This memoir develops the whole basis of the theory of elasticity; most importantly, strains are properly analysed and related in terms of displacements of the material, and Saint-Venant obtains the well-known solutions for torsion of bars of, for example, elliptical, rectangular and triangular cross-section. Since he has the equations at hand, he applies them also to Galileo's problem, and determines the distribution of shear stress over the cross-section. These bending solutions are developed in more detail in the second memoir, 1856.

In editing Navier's text in 1864, Saint-Venant deals first with the question of strain. Shear strain is identified as an angular distortion, and Saint-Venant gives the proper value for principal strain at a point where the (two-dimensional) strains are e_{xx}, e_{yy} and γ_{xy} as

$$e_{\max} = \frac{1}{2}\left\{(e_{xx} + e_{yy}) + \left[(e_{xx} - e_{yy})^2 + \gamma_{xy}^2\right]^{\frac{1}{2}}\right\}. \tag{3.4}$$

He then applies this to the problem tackled by Navier, which led to equation (3.1), or, in non-dimensional form,

$$m^2 + p^2 = 1. \tag{3.5}$$

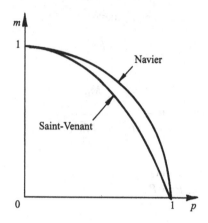

Fig. 3.1. Interaction diagram for moment of resistance (m) in the presence of shear force (p).

As a numerical example for a particular value of Poisson's ratio, Saint-Venant replaces this circular criterion by an expression which leads to the elliptical criterion

$$\frac{1}{4}m^2 + \frac{3}{4}m + p^2 = 1; \tag{3.6}$$

the two criteria are sketched in fig. 3.1.

In addition to his faulty analysis of strain Navier had, in effect, assumed a uniform distribution of shear strain across the whole depth of the cross-section of the beam. Up to this point in his commentary, Saint-Venant takes the same uniform distribution of shear strain, and is concerned to give the proper analysis of combined strains. Some idea of the scope of the 'footnotes' to Navier's text is given by Table 3.1. Paragraph 156 of Navier's text starts a new section, dealing with torsion, and Saint-Venant's footnotes extend to 264 pages. He presents and develops the whole theory of torsion of non-circular members given in his 1855 memoir; some 150 pages into these notes, he comes back to Galileo's problem, and he determines the distribution of shear stress at the root of the cantilever. In his analysis he follows the approximate theory of Jouravski (1856), in which an imaginary horizontal cut is made at the cross-section at which a shear force P is acting, fig. 3.2. If it is assumed (and this is one of the approximations) that the shear stress acts uniformly on this horizontal

Table 3.1. *Pages devoted by Navier and by*
Saint-Venant to paragraphs 151–156 of the Leçons

Paragraph no.	151	152	153	154	155	156
Navier	1	2	1	1	(6 lines)	2
Saint-Venant	12	19	9	21	–	264

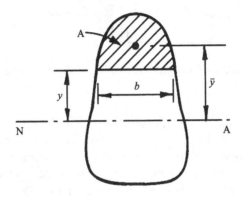

Fig. 3.2. Imaginary cut in cross-section for analysis of shear stress.

cut, then its magnitude τ is given by

$$\tau = \frac{P A \bar{y}}{I b},\tag{3.7}$$

where I is the second moment of area of the cross-section, and $A\bar{y}$ the first moment of area of the shaded section in fig. 3.2, all referred to the neutral axis of elastic bending. From this Saint-Venant easily shows that the elastic shear-stress distribution over a rectangular cross-section is parabolic, with a maximum value (at the neutral axis) of $\frac{3}{2}$ times the 'mean' shear stress.

Because of his work on exact solutions of the governing equations, Saint-Venant is well aware of the approximate nature of equation (3.7). He notes that it will only be somewhere near the truth for non-rectangular sections if the section is thin (that is, if the dimension b in fig. 3.2 is small compared with the overall depth); the *total* force acting on the horizontal cut will be evaluated correctly, but the stresses will not be distributed uniformly over the cut. This restriction on the interpretation of equation (3.7) was not always appreciated by later analysts.

The Effect of Shear

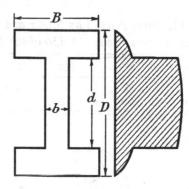

Fig. 3.3. The incorrect ('top-hat') distribution of shear stress in an I-section.

Saint-Venant applies the approximate elastic theory to the I-section, and notes that it is the web that carries by far the greatest part of the shear force P, while the flanges carry the bending moment M. Moreover the shear stress in the web, while varying parabolically, in fact varies only slightly. This observation has led to the simple and effective design rule that the shear stress in an I-section beam is merely the value of the shear force divided by the area of the web; the effect of shear on the flanges can be neglected. (The unthinking transfer of this design procedure from the I-section to the box-section girder proved dangerous for the design of steel bridges of this type; see §3.3 below.)

Jouravski's theory, accompanied by sometimes unremarked approximations, quickly passed into the standard texts. In built-up I-sections, for example, in which the flanges are attached through angles to the web by rivets or bolts, the theory gives a quick way of designing those bolts. In the same way the theory may be used in elastic composite design to determine the size and spacing of shear connectors joining a concrete slab to a steel joist. However, Saint-Venant's reservation about non-uniform distribution of shear stress was not always remembered. Ewing (1899), for example, remarks that 'the intensity of shearing stress is nearly uniform over the web of an I-section, and is much greater there than in the flanges in consequence of the much smaller value of the width [of the web]'.

Case (1925, and in later editions, 1932 and 1938), realises that a horizontal cut through the flange of an I-beam is not the correct approach, but he gives no solution to the problem; indeed, he states that the stresses in the flanges 'are not open to calculation'. Chilver corrects this in the post-war edition, Case and Chilver (1959). Even Timoshenko himself, in the fourth edition (1962) of his *Elements of strength of materials* (1935), echoes Case when he states that the question of stresses in flanges 'is too complex to be analyzed by elementary methods'. As late as 1968 Pippard and Baker's text (fourth edition) illustrates the incorrect 'top-hat' distribution, fig. 3.3, in which the shear stress in the flanges is calculated as a mean value over the full width B. By contrast Shanley (1957), for example, not only deals properly with the I-section but also applies correct theory to the hollow box section.

3.3 Thin-walled sections

Designers working in the field of civil engineering structures are not usually involved with the use of thin-walled members; the flange of a steel I-beam, for example, while not having the relatively gross dimensions sketched in fig. 3.3, is nevertheless sufficiently thick that the magnitude of the shear stress is not a critical design consideration. The fact that the shear stress distribution sketched in fig. 3.3 is wrong is of no practical consequence. Aircraft designers, however, were using thin-walled members, and Timoshenko studied some of the problems in the 1920s and 30s; the work was taken up by Vlasov (1940) who gave a comprehensive account.

The essential feature of the behaviour of a thin-walled member in shear is that the shear stresses act in a direction parallel to the surfaces of the cross-section. Thus for the channel section the cuts should be made as shown in figs 3.4(*a*) and (*b*); equilibrium of forces acting on the shaded sections, for which bending stresses are calculated by simple elastic theory (just as for the general section of fig. 3.2 leading to equation (3.7)), will then give values for the corresponding shear stresses. These shear stresses are shown schematically in the sketch of fig. 3.4(*c*).

Similarly, fig. 3.5 illustrates the correct elastic shear-stress distribution for the I-section, and fig. 3.6 that for the box section. A study of the statics of fig. 3.6 reveals that the flanges of a box section can be subject to high transverse compressive forces, and they must be designed to resist buckling.

3.4 Plastic solutions

The analysis of shear stress leading, for example, to the sketch of fig. 3.4(*c*), relies on a knowledge of the distribution of bending stress

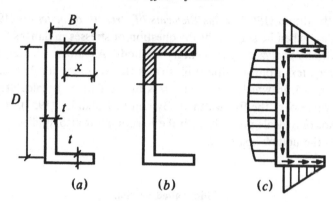

Fig. 3.4. Elastic shear stresses in a channel section. (*a*) and (*b*): cuts corresponding to fig. 3.2; (*c*) shear stresses are linear for the flanges, parabolic for the web.

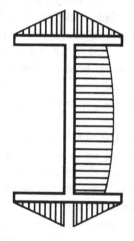

Fig. 3.5. Elastic shear-stress distribution for the I-section.

at the cross-section. Thus if a linearly varying distribution of bending stress is introduced into the equations (of equilibrium), then all the approximations entailed in Navier's simple elastic bending theory are carried through to the calculation of shear stress. The procedure allows the evaluation of stresses at a local section (the root of the cantilever in Galileo's problem); some criterion of failure (for example, that of Navier, i.e. maximum strain) is then applied to give an indication of the critical condition.

From Saint-Venant's commentary, it may be inferred that he was well

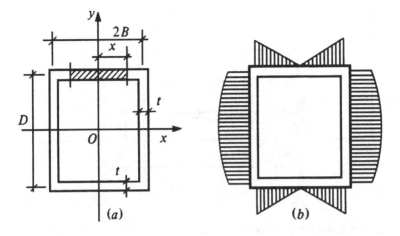

Fig. 3.6. Elastic shear-stress distribution for the box section.

aware that such a 'solution' to Galileo's problem was incomplete. By inspection, it is reasonable to assume that the root of the cantilever is the most critical section, but examination of that section alone is insufficient; a full analysis requires the evaluation of stresses throughout the cantilever. As will be seen, this observation is not trivial or pedantic. (In passing, it may be repeated that Galileo stated his problem as that of finding the value of the tip load that would break a cantilever, but that he himself transformed this problem of the theory of structures into one of strength of materials – the evaluation of the moment of resistance at the root of the cantilever.)

From the engineering point of view, however, what is needed is an investigation of the effect of shear force on the value of the moment of resistance at the root of Galileo's cantilever. Those concerned with the development of plastic theory were aware that shear stress would have an effect on the value of the full plastic moment, in some way similar to the effect of axial load discussed at the end of the previous chapter. Given the imprecise thinking about shear stress as late as the 1940s and 50s, it is not surprising that the problem proved difficult.

It is indeed difficult, since the analysis must be three-dimensional, and not confined to the cross-section at the root of the cantilever. Some experimental evidence was available, and it seemed that the plastic hinge formed at collapse of the cantilever involved outer plastic zones failing in pure tension and compression, and a central 'shear' zone, in which failure

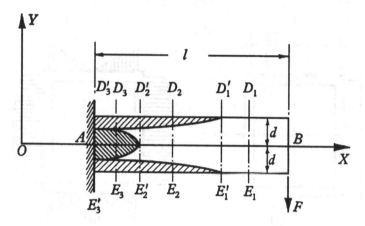

Fig. 3.7. Collapse of a rectangular cross-section cantilever, from Baker *et al.* (1956).

was actually under combined shear and bending stresses. Figure 3.7, from Baker *et al.* (1956), illustrates these zones, sketched for convenience for the rectangular cross-section.

At a section such as D_2E_2 in fig. 3.7 the outer zones may be shown to be yielding in pure tensile or compressive stress, as sketched in fig. 3.8, the shear force being carried by shear stresses in the elastic central core. The usual simple theory gives the expected distributions in those portions of the cross-section that remain elastic. For sections such as D_3E_3 in fig. 3.7, a study of the equilibrium equations enables the extent of the inner plastic zone to be determined. The analysis was made by Horne (1951), and this type of approach gives, according to plasticity theory (see Chapter 9), a 'safe' lower-bound estimate of the collapse load of the cantilever. Horne's solutions were incomplete, but the lower-bound estimate was good. Neal (1961*a*, *b*) gave full solutions leading to true lower bounds; these are discussed below.

Horne's analysis can be applied without change to I-section beams, provided that the flanges are fully plastic, and fig. 3.9 compares the reduction in carrying capacity of a beam of rectangular cross-section with that of an 8 in × 4 in I-section. It will be seen that, as confidently assumed by Coulomb and Navier, it is only the very shortest beams for which, in an engineering sense, any significant effect may be noted.

For the engineering design of the I-beam, Heyman and Dutton (1954) proposed an empirical approach based upon the practical approximation already noted, namely that the shear stress may be considered to be

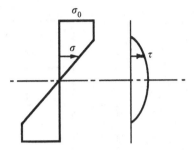

Fig. 3.8. Stress distributions at section D_2E_2 of the cantilever of fig. 3.7.

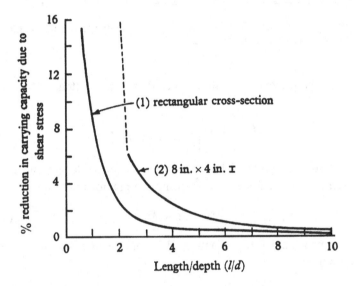

Fig. 3.9. Reduction in carrying capacity due to shear stress.

uniformly distributed over the web. At full plasticity, the assumed stress distributions are as sketched in fig. 3.10, with the flanges fully stressed to the yield stress σ_0, and the web subjected to combined stresses σ and τ, where

$$\sigma^2 + k\tau^2 = \sigma_0^2, \qquad (3.8)$$

and the constant k may be either 3 or 4 according as the von Mises or the Tresca yield condition is taken. In fact, for any value of k, if the maximum allowable shear force sustainable by the web is denoted P_0, then the moment of resistance corresponding to the distribution of

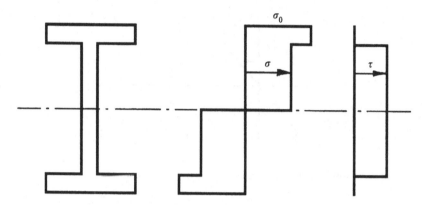

Fig. 3.10. Approximate stress distributions corresponding to full plastic moment formed in the presence of shear force.

fig. 3.10 in the presence of a shear force P may be written

$$M = M_f + \left[1 - \left(\frac{P}{P_0}\right)^2\right]^{\frac{1}{2}} M_w, \tag{3.9}$$

where M_f and M_w are the full plastic moments of the flanges and web in the absence of shear force, so that, using the previous notation,

$$M_0 = M_f + M_w. \tag{3.10}$$

In non-dimensional form, equation (3.9) for the moment of resistance may be written

$$m = m_f + \left(1 - p^2\right)^{\frac{1}{2}} m_w, \tag{3.11}$$

and this curve is sketched in fig. 3.11. This interaction diagram may be compared with those of fig. 3.1 due to Navier and Saint-Venant.

Curves such as those shown in figs 3.1 and 3.11 are not true yield surfaces in the sense required by plasticity theory (Heyman (1970)). Neal presents both upper and lower bounds to the solution to the problem, and curves for the first quadrant are sketched in fig. 3.12. For the upper bounds, Neal used velocity fields similar to those proposed by Leth (1954); Green (1954) also evaluated upper bounds. If it were possible to construct a proper yield surface for the variables M and P, the bending moment and shear force acting at a cross-section, then that yield surface should be convex; it is at once apparent that no convex curve can

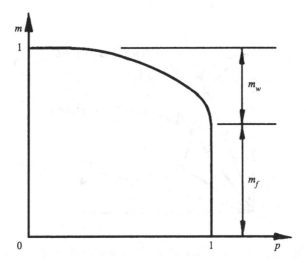

Fig. 3.11. Approximate interaction diagram for the full plastic moment of an I-section in the presence of shear force.

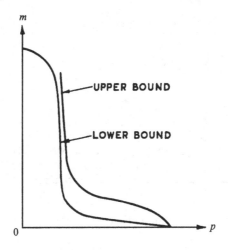

Fig. 3.12. Interaction diagram for I-section (after Neal).

be constructed to lie between the upper and lower bounds marked in fig. 3.12.

The fact is that Galileo's cantilever has only one loading parameter, the tip load P; the bending moment M is, in effect, the same variable, since $M = P\ell$. The exact solution of the plastic collapse problem for Galileo's cantilever is difficult, but it is certainly possible to determine upper and

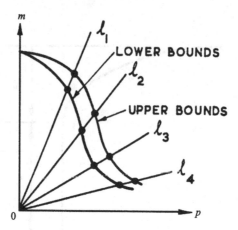

Fig. 3.13. Bounds on the interaction diagram for an I-section.

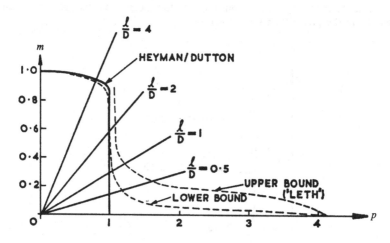

Fig. 3.14. Empirical curve for an I-section compared with Neal's bounds.

lower bounds for the value of the tip load for a given cantilever of length
ℓ. These bounds could be recorded on an m/p diagram, fig. 3.13, for a
cantilever of length ℓ_1; similar points could be plotted for a cantilever
of length ℓ_2, and so on, and the points connected to give an interaction
diagram. This diagram is not, as has been mentioned, a proper yield
surface.

Neal, following Leth, made calculations on the US rolled I-section
8WF40. The results are shown in fig. 3.14 with the approximate curve

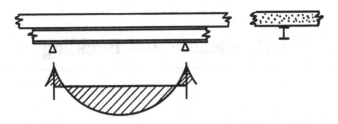

Fig. 3.15. Typical bending moments for a continuous composite beam.

of equation (3.11) (fig. 3.11) superimposed. The radial lines represent different length/depth ratios of the cantilever; steel beams for which $\ell/D < 2$ are not of much practical importance, although in the composite steel/concrete construction of continuous beams it is possible for points of contraflexure to lie quite close to supports, giving an effectively short 'Galilean' cantilever, fig. 3.15.

4

Elastic Flexure and Buckling

It has been noted that James Bernoulli (1694, 1695) discussed the problem of finding the moment of resistance of a cross-section in bending. This same paper makes a fundamental contribution to the problem of the elastic flexure of a member. Bernoulli remarks that Galileo had contended (wrongly) that the deflected form of the cantilever was a parabola. Saint-Venant, in his annotated edition of Navier's *Leçons*, repeats this attribution to Galileo, but in fact there is no such contention to be found in the *Dialogues* of 1638. The first discussion of an elastic deflected form seems to be that of Pardies (1673), and he indeed asserts that the parabolic form is correct.

Pardies starts his book on *Statics* with clear and accurate statements of basic laws – the law of the lever, for example (in which his presentation follows exactly that of Galileo, fig. 1.2), moments of forces, the laws of pulleys, the forces in windlasses, gear trains and so on. He then moves on to discuss the question of the shape of a hanging uniform cord, and he establishes the powerful 'Pardies' theorem', namely that the tangents at any two points on the cord intersect at a point directly below the centre of gravity of the portion of the cord between the two points. He states that the shape of the hanging cord is *not* a parabola, and settles finally for the hyperbola (he had, of course no knowledge of the calculus. Leibniz (1691) published the solution of the catenary).

Extraordinarily, however, Pardies states, and in a muddled way proves, that if a weightless hanging cord is subjected to loading that is uniformly distributed *horizontally* (instead of along the cord) then the shape will indeed be that of a parabola.

Pardies also extends Galileo's discussion of the cantilever beam of equal resistance. He notes Galileo's design – a beam of uniform width and parabolic depth – and adds the design of a section of uniform depth and linearly tapering width. For Pardies, fracture occurs because of extension – or, to use modern terminology, the attainment of a maximum tensile strain. As an example of the application of this criterion, Pardies has a brief section on the difficulty of breaking an egg (in the hand) by applying pressure from end to end. Pardies' argument is, typically, unconvincing and non-mathematical, although the observation is exact; pressure along the axis tends not to bend the shell of the egg, and only bending will produce tensile strain leading to fracture. By contrast, pressure along the smallest diameter of the shell will engender bending.

4.1 James and Daniel Bernoulli

James Bernoulli (1691) made the first analytical contribution to the problem of elastic flexure of a beam. He published a logogriph: *Qrzumu bapt dxqopddbbp* ..., whose secret he revealed in 1694; a letter is replaced by the next in the Latin alphabet, the second by the letter three away, and the third by the letter six away, so that *aaaaa* ... would be encoded *bdgbd* ... The logogriph thus reads 'Portio axis applicatam ...', and the decoded statement is that the radius of curvature at any point of an initially straight beam is inversely proportional to the value of the bending moment at that point.

All this was amplified in Bernoulli's *Explicationes* of 1695, in which he considers the relative inclinations of two neighbouring cross-sections of the beam. The neutral axis is taken at the face of the section, but this does not affect the result that curvature is proportional to bending moment. Bernoulli does not make the assumptions of small slopes, and hence the differential equation for the elastic curve is not soluble in terms of elementary functions. However, his equation is, after manipulation, of first order only and hence easily soluble in terms of a series expansion. (Saint-Venant shows that, by a suitable choice of axes, the general second-order differential equation of bending can always be integrated once without approximations.)

James Bernoulli, then, was the first to give a solution to the problem of the shape of a bent elastic strip of uniform cross-section – the problem of the elastica. The mathematical difficulties lie in the fact that the general expression for curvature referred to rectangular axes is

$$\frac{\dfrac{d^2y}{dx^2}}{\left[1 + \left(\dfrac{dy}{dx}\right)^2\right]^{\frac{3}{2}}}, \tag{4.1}$$

or (what is the same thing)

$$\frac{d^2y}{dx^2} \bigg/ \left(\frac{ds}{dx}\right)^3, \tag{4.2}$$

where s is length measured along the strip. If the bending moment is taken as a linear function of x (Galileo's problem of a tip-loaded cantilever), then Bernoulli's first integration gives

$$\frac{dy}{dx} = \frac{x^2}{\left(c^4 - x^4\right)^{\frac{1}{2}}}, \tag{4.3}$$

where c is a constant involving the tip load and the flexural rigidity of the cantilever. Bernoulli integrated equation (4.3) in terms of infinite series, and obtained approximate answers (in fact, very close bounds) by evaluating those series numerically.

As opposed to an approximation to the *solution* to the bending problem, Daniel Bernoulli (1741–43), nephew of James, seems to have been the first to propose the 'engineering' approach, in which the approximation is made to the fundamental equation rather than to the exact solution. Daniel Bernoulli saw that if deflexions were small and if the x-axis was taken along the length of the cantilever, then $ds \approx dx$ in equation (4.2), so that the curvature was now simply d^2y/dx^2, and the equation for Galileo's cantilever was

$$EI\frac{d^2y}{dx^2} = Wx. \tag{4.4}$$

Daniel Bernoulli wrote this equation, which is simply integrable in quadratures, and obtained the proper cubic equation for the deflected form and hence the value of the tip deflexion under a load W. His work is actually a preliminary to a deeper and more difficult study, namely that of determining the modes and frequencies of vibration of a uniform cantilever. (He is aware from his solutions that overtone frequencies of the cantilever are not simple multiples of the fundamental, but are irrational with respect to the fundamental; dissonance of a tuning-fork is inevitable.)

4.2 Euler and the elastica

Saint-Venant states that it was Navier who first formulated clearly the fact that small-deflexion theory of elastic bending could always be made to lead to simple equations, such as equation (4.4), which were easily integrable. All that is necessary is to take the x-axis as tangent to some point of the (deflected) beam; $(dy/dx)^2$ can then be neglected compared with unity.

Euler did not locate his axes in this way; usually he oriented them with respect to the applied loading. In any case, his great work on the problem of the elastica does not make the assumption of small deflexions. The problem had been under lively discussion since the time of James Bernoulli, but it was Daniel who wrote to Euler in 1742 suggesting his intervention (in fact, effectively challenging him to intervene). Daniel Bernoulli had found that the *vis viva potentialis laminae elasticae*, $\int ds/R^2$, where R is the radius of curvature, was a minimum for the elastic curves of his uncle James (that is, the strain energy in bending was a minimum). He proposed that Euler should apply his calculus of variations to the inverse problem of finding the shape of the curve of given length so that the function $\int ds/R^2$ was a minimum. The two endpoints of the curve were to have specified positions and slopes, and the strip was to be unloaded except at those ends.

Euler (1744) makes the analysis in the '*Additamentum I, De Curvis Elasticis*' to his book *Methodus inveniendi lineas curvas ...*, which is, in fact, a book on the calculus of variations. Following Daniel Bernoulli's suggestion, Euler immediately obtains the fourth-order differential equation, and then, with the utmost brilliance, integrates it three times to obtain

$$\frac{dy}{dx} = \frac{\left(a^2 - c^2 + x^2\right)}{\left(c^2 - x^2\right)^{\frac{1}{2}} \left(2a^2 - c^2 + x^2\right)^{\frac{1}{2}}}, \tag{4.5}$$

(cf. James Bernoulli's equation (4.3) for Galileo's cantilever). The derivation has been purely mathematical, but Euler then shows that equation (4.5) can be derived directly (without using the calculus of variations) from James Bernoulli's basic statement that moment is proportional to curvature. With this physical interpretation, the external load (applied through the ends) can be reduced to a single force P, and $a^2 = 2EI/P$, where EI is the flexural rigidity of the elastic strip.

The constant c in equation (4.5) is a constant of integration, and the form of the solution of the equation is strongly governed by the relative

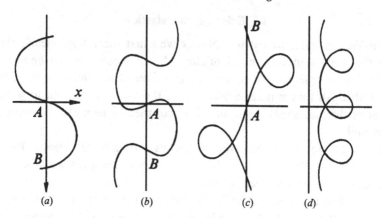

Fig. 4.1. Elastic curves of the second, fourth, sixth and eighth classes (after Euler (1744)).

magnitudes of a and c. At the origin, the slope of the elastica is

$$\left[\frac{dy}{dx}\right]_{x=0} = \frac{a^2 - c^2}{c\left(2a^2 - c^2\right)^{\frac{1}{2}}}, \qquad (4.6)$$

and for $a = c$, for example, this slope is zero, while for $c^2 > 2a^2$ the curve cannot cut the y-axis at all. Euler proceeds, with very little calculation at this stage, to distinguish nine classes of solution of equation (4.5), of which his classes 2, 4, 6 and 8 are sketched in fig. 4.1. Classes 3, 5 and 7 are unique solutions representing transition curves between those sketched in fig. 4.1, while classes 1 and 9 represent the endpoints of the analysis.

For c very small, for example, the deflexions are of the pattern of fig. 4.1(a), but those deflexions are also very small. For $c = 0$ Euler shows easily that the infinitesimal deflexions are sinusoidal, and that they can only be maintained for a certain finite value of the force P, equal to (the 'Euler buckling load') $\pi^2 EI/\ell^2$. (Euler imagines a string AB connected to the elastica; the string must of necessity carry this finite force.)

For the general second class, fig. 4.1(a), the force necessary to maintain equilibrium is greater than $\pi^2 EI/\ell^2$. The class arises for $0 < c < a$, and, from equation (4.5), it is clear that the curve is confined to lie between $x = \pm c$. Class 3 is the special case $c = a$, and class 4 occurs for $a < c^2/a^2 < 1.651868$, fig. 4.1(b). At the upper limit, class 5, points A and B coincide, and the curve has the form of a figure 8; for larger values of c, the general shape is as shown in fig. 4.1(c). At $c^2 = 2a^2$, class 7, the curve becomes asymptotic to the *y*-axis, and for class 8 it lies wholly to the right of that axis, fig. 4.1(*d*). Finally, for *c* very large, class 9, the elastica is bent into a portion of a circle. All these curves represent the elliptic integrals which arise from the basic equation, and Euler obtains solutions in power series of c^2/a^2, from which he calculates numerical results (to a large number of significant figures).

It is, of course, the fundamental result of class 1, that of the Euler buckling load, that has been of importance for the structural design of columns. Euler himself was fully aware of this importance, and he gave specifically the (Euler) buckling load for a pin-ended column. He returned to the matter in his Berlin Mémoire of 1757, in which, in the manner of Daniel Bernoulli, he approximated the governing equation rather than its solution, and wrote directly (for class 1, small deflexions) the familiar engineering equation

$$EI\frac{d^2y}{dx^2} = -Py. \qquad (4.7)$$

Euler did not explore the higher modes which are implicit in equation (4.7). It was Lagrange (1770–73) who showed that the buckling load of a pin-ended column could be expressed as $m^2\pi^2 EI/\ell^2$, where *m* is any integer; his discussion is along the lines that are followed by a standard elementary text of today.

4.3 Buckling of columns: the nineteenth century

Euler's basic result, that buckling of a column is inversely proportional to the square of the length of the column, had been established experimentally somewhat earlier. As mentioned in Chapter 2, Musschenbroek (1729) had published the results of a very large number of material tests, mainly on wood, using his own designs of testing machines to apply tension, compression and bending. The bending tests confirm Galileo's result that the ultimate strength of rectangular beams is proportional to the width and to the square of the depth of the cross-section. The compression tests are the first recorded for struts.

Coulomb (1773) specifically rejects Musschenbroek's formula (and, therefore, Euler's also, although he does not mention Euler's work on buckling); Coulomb's own tests, on masonry columns, indicated that the breaking strength was independent of length. Indeed, stocky stone columns do not buckle; their failure load, as Coulomb found experimentally, and for which he provided theory, is merely proportional to cross-sectional area.

The introduction of iron into structural design prompted tests on that material. For example, Duleau (1820) carried out bending tests, and also made compression tests on slender struts, confirming the form of expression for Euler's buckling load. A little later, Hodgkinson (1840) reported similar tests, and he also confirmed that Euler's formula could be applied to slender columns. Moreover, his experiments covered satisfactorily 'pinned' and 'fixed' end conditions, and he deduced that the buckling loads of a pin-ended column and of a fixed-ended column of twice the length were the same. However, shorter columns no longer obeyed Euler's rule, and Hodgkinson established empirical formulae for such columns, which were used for some time in the nineteenth century as the basis for design. It was clear, of course, that short columns would in theory buckle at very high stresses if Euler's formula were to be applied without modification – stresses larger than any possible failure stress of the material.

Lamarle noted this. The second part of his Mémoire (1846) deals with axial loading; he transforms Euler's formula into terms of a critical stress σ_{cr}:

$$\sigma_{cr} = \frac{\pi^2 E}{(\ell/r)^2}, \tag{4.8}$$

where r is the radius of gyration (about the weaker principal axis) of the cross-section, so that $I = Ar^2$, and ℓ/r is the 'slenderness ratio' of the (pin-ended) strut. Lamarle proposed that Euler's formula should be used provided that σ_{cr} were less than the elastic limit σ_0 for the material, that is, for a strut whose slenderness ratio was greater than the value given by

$$\left(\frac{\ell}{r}\right)^2 = \frac{\pi^2 E}{\sigma_0}. \tag{4.9}$$

For shorter struts, the maximum stress that could be carried would be σ_0, and the strut 'design curve' of fig. 4.2 emerges.

There is no record of Lamarle's proposal having been used in practice, but Gordon's empirical formula, as given in Rankine's handbook of the middle of the nineteenth century (1862 and many later editions through to the twentieth century), gives the same stresses as Lamarle for small and large slenderness ratios. The strut is imagined to bend, so that the maximum stress (always for a pin-ended strut) results from

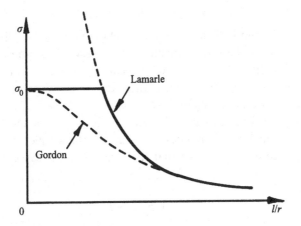

Fig. 4.2. Design curves for struts.

the superimposition of a direct compressive stress and a stress due to bending. If this maximum stress is set equal to σ_0, then the empirical formula is determined as

$$\sigma_0 = \frac{P}{A}\left[1 + a\left(\frac{\ell}{r}\right)^2\right],\qquad(4.10)$$

so that

$$\sigma_{cr} = \frac{\sigma_0}{1 + a\left(\dfrac{\ell}{r}\right)^2}.\qquad(4.11)$$

Gordon's curve, equation (4.11), is sketched in fig. 4.2. Gordon (as quoted by Rankine) gave proposals for values of σ_0 and the constant a for various materials (wrought iron, cast iron, timber) making use of Hodgkinson's experimental results, and allowance was made for both fixed and pinned ends. For example, for a pin-ended wrought-iron strut,

$$\sigma_{cr} = \frac{36\,000}{1 + \dfrac{1}{9000}\left(\dfrac{\ell}{r}\right)^2}.\qquad(4.12)$$

where the stress is measured in lb/in^2. (For a fixed-ended column of the same material, the constant in the denominator is $36\,000$ rather than 9000.)

4.4 Perry–Robertson

The Gordon/Rankine formula, still embedded in some building codes, was abandoned in the UK in favour of the Perry–Robertson approach. Ayrton and Perry (1886) analysed a centrally loaded column which has some initial curvature, giving an initial small displacement a at mid-height. If the initial shape is sinusoidal (or, indeed, if attention is confined to the first term of a Fourier expansion for a more general shape, since that term dominates the analysis) the simple theory (i.e. a modified equation (4.7)) shows that the deflexion a increases under end load P to

$$a' = a \left(\frac{P_E}{P_E - P} \right), \tag{4.13}$$

where P_E is the Euler buckling load of the pin-ended strut, or, what is the same thing,

$$a' = a \left(\frac{\sigma_E}{\sigma_E - \sigma} \right). \tag{4.14}$$

The total maximum compressive stress at mid-height is therefore made up of an axial stress σ together with a bending stress due to the moment Pa'; that is

$$\sigma + \frac{Pac}{I} \left(\frac{\sigma_E}{\sigma_E - \sigma} \right), \tag{4.15}$$

where c is the distance from the neutral axis of the section to the outermost compression fibre. Setting $I = Ar^2$, and assuming that failure of the strut occurs when the total stress reaches the elastic limit σ_0, expression (4.15) gives

$$\sigma_0 = \sigma \left(1 + \frac{ac}{r^2} \frac{\sigma_E}{\sigma_E - \sigma} \right). \tag{4.16}$$

This is a quadratic for the value of the critical stress σ, viz.

$$\sigma^2 - \sigma \left[\sigma_0 + (1 + \eta)\sigma_E \right] + \sigma_0 \sigma_E = 0, \tag{4.17}$$

where $\eta = ac/r^2$. Equation (4.17) is the Perry formula from which, if the value of η is known, the value of critical stress that will cause failure of the strut may be calculated.

For struts of similar geometrical cross-section, the value of c is proportional to that of r, so that the value of η may be written

$$\eta = \frac{ac}{r^2} \propto \frac{a}{r} = \frac{a}{\ell}\lambda, \tag{4.18}$$

where λ is the slenderness ratio ℓ/r. Robertson (1925) made his own extensive theoretical and experimental investigation of the strut problem, and he reviewed the work of others. He concluded that a/ℓ was roughly constant for practical columns used in building – the crookedness of a strut was proportional to its length. An average value of η was $(10^{-3})\,\lambda$, and the worst (most conservative) value $(3 \times 10^{-3})\,\lambda$. This last value is the one used in compiling the tables for the 1969 British Standard for steel design (or rather, for an unexplained reason, the value used is $0.3(\lambda/100)^2$. With the particular numbers chosen, both the linear and the quadratic form give the same value of σ from equation (4.17) for $\lambda = 0$ and 100). The general solution of equation (4.17) is of the Gordon/Rankine form sketched in fig. 4.2.

(For the purposes of practical design, the Steel Structures Research Committee, whose work will be referred to again in Chapter 9, recommended in the 1930s that a safety factor on stress of value 2.36 should be introduced. That is, for a certain value of σ calculated from equation (4.17), the maximum permitted stress should be 1/2.36 of this value. The figure of 2.36 was reduced in later British building codes, to 2 and then to 1.7, before being replaced by the concept of a factor on the applied loads rather than a safety factor on stress.)

It will have been noted that all of this work directed to the solution of the problem of design of a column in a practical building has concentrated on the behaviour of a pin-ended strut (although, from Rankine onwards, it was always stated that a column with 'flat ends' has four times the buckling load of the corresponding pin-ended member). The formulae, combining as they do an 'Euler' flexural analysis with the idea of a limiting 'squash load', have introduced empirical constants, but two major modifications are necessary before they can be applied to a practical structure. Firstly, neither the pinned nor the fixed end represents reality; a practical column will be connected to other members which may provide both positional and flexural restraint. To allow for different end conditions the idea of 'effective length' is used – the idea can in fact be traced back to Euler in 1759. Thus a fixed-ended column buckles at four times the load for a pin-ended column and, since Euler buckling depends on the inverse square of the length, the effective length of a fixed-ended column is one-half of that of the same member with pinned ends. Similarly, Galileo's cantilever under axial load has an effective length of twice that of the pin-ended column. The designer working to a code will assess an effective length of a particular column in a real construction and use that value in the formulae.

Secondly, the practical connexions to a column, instead of restraining the ends, may induce bending – or, indeed, the column may be part of a frame and be required to contribute bending strength as well as to carry axial load. For the purposes of a simple practical design method, a number of codes adopt a kind of Perry approach; that is, the stress due to axial load and the stress due to applied bending moment must together not exceed a certain limit. The code referred to above (British Standard 449: Part 2: 1969 (The use of structural steel in building) with amendments through to 1989) gives a formula

$$\frac{f_c}{p_c} + \frac{f_{bc}}{p_{bc}} < 1, \tag{4.19}$$

where f_c is the calculated average compressive stress,

         p_c is the allowable compressive stress in axially-loaded struts (i.e. the value of σ which results from equation (4.17)),

         f_{bc} is the resultant compressive stress due to bending about both rectangular axes

and p_{bc} is the allowable compressive stress for members subject to bending.

The interaction diagram implied by inequality (4.19) is sketched in fig. 4.3; a 'safe' design is supposed to lie on the origin side of the straight line. (As noted below, British Standard 5950: Part I: 1990 (Structural use of steelwork in building), while still making use of Perry–Robertson ideas, has a more sophisticated approach to the design of columns.)

4.5 Lateral-torsional buckling

The 'allowable compressive stress' in bending, p_{bc}, which appears in inequality (4.19), is not a constant, but is a function of the slenderness ratio of the column about its minor axis. The eighteenth-century discussion of the flexure of the elastica was concerned with a member of uniform cross-section, perhaps assumed to be rectangular but in fact not usually specified. However, implicit in these analyses was that bending was taking place about what is now called the minor axis of the cross-section, and it has been seen that Navier (1826), or more clearly Saint-Venant (1864), discussed the general case of such flexure. No mention was made at that time of the possible instability of a beam when it was bent about the major axis. It seems that in 1899 both Prandtl and Michell, working

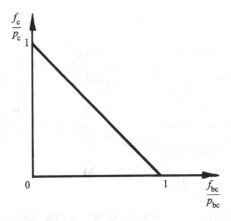

Fig. 4.3. Design chart for column subjected to combined bending and axial load (BS449: 1969).

independently, found solutions to the problem of the buckling of a thin deep rectangular cross-section bent about the major axis.

Timoshenko (1936) discusses Prandtl's work. Figure 4.4 is reproduced from Timoshenko's text on stability, in which he gives a simple account of the lateral buckling, accompanied by twisting, of a thin deep rectangular strip under pure bending. The equation governing the angle of twist ϕ is

$$C\frac{d^2\phi}{dz^2} + \frac{M^2}{EI}\phi = 0 \qquad (4.20)$$

where M is the value of the applied moment, and EI and C are, respectively, the flexural rigidity about the minor axis and the torsional rigidity of the cross-section. (The value of I is of course $hb^3/12$. The value of C is approximately $\frac{1}{3}hb^3G$, where G is the shear modulus of the material, but see below for a discussion of 'Saint-Venant' and warping rigidities.)

Equation (4.20) is exactly analogous to Euler's 'engineering' equation for buckling under axial compression; it has the trivial solution $\phi = 0$ unless

$$M = \frac{\pi}{\ell}\sqrt{EIC}, \qquad (4.21)$$

in which case ϕ varies sinusoidally and has indeterminate amplitude (which are, of course, characteristics of the Euler problem).

The value of M given by equation (4.21) gives the critical condition for lateral-torsional buckling of a rectangular beam in pure bending. The analysis may be repeated for the cantilever under end load W (Galileo's

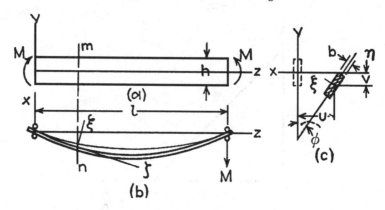

Fig. 4.4. Thin deep rectangular strip buckling under the application of equal end couples M (from Timoshenko (1936)).

problem) – the beam buckles out of plane with combined bending and twisting at a load determined by the equation

$$C\frac{d^2\phi}{dz^2} + \frac{W^2}{EI}(\ell - z)^2 \phi = 0, \tag{4.22}$$

where the origin of co-ordinates is taken at the left-hand (fixed) end of the cantilever. According to Timoshenko, Prandtl (1899) solved equation (4.22) in terms of infinite series, which he summed numerically to find the critical load W_{cr} (for which a solution exists other than the trivial $\phi = 0$). In fact, as noted by Michell, the solution of equation (4.22) can be obtained in terms of Bessel functions of order $\frac{1}{4}$ and $-\frac{1}{4}$, and the equation giving the critical load is

$$J_{-\frac{1}{4}}\left(\frac{W\ell^2}{2\sqrt{EIC}}\right) = 0, \tag{4.23}$$

of which the lowest root is

$$W_{cr} = \frac{4.013}{\ell^2}\sqrt{EIC}. \tag{4.24}$$

These and other solutions are discussed by Timoshenko (1936) and in the second edition of the same text by Timoshenko and Gere (1961). The solutions apply to thin deep beams in which warping of the cross-section is of little consequence.

For the more general cross-section, a beam twisted by end couples will tend to warp – an initially plane cross-section does not remain plane, but will distort to follow the helical pattern of the twisting deformation. If this warping is not restrained in any way, then the torsional rigidity C (in equations (4.20) and (4.21) for example) is the 'Saint-Venant' rigidity; Saint-Venant (1855) gave the first correct examination of torsion under these conditions. Timoshenko (1905, 1906) realised that if warping were prevented (by the ends of the beam being constrained to remain plane, for example, or, in general, by the interaction of neighbouring cross-sections), then a bending deformation of the cross-section would be introduced that would increase the resistance of the beam to twisting.

Timoshenko examined the problem of torsion of an I-beam; he did not establish a general theory, but calculated an extra torsional rigidity C_1 by estimating the bending contributions of the flanges of the beam. The 'warping rigidity' C_1 is now sometimes called the Vlasov rigidity; Vlasov (1940) gives a full and connected account of the torsion of thin-walled bars of any general cross-section, and shows how the rigidity C_1 should be calculated. However, Timoshenko (1905) analysed the general problem of torsion correctly, and deduced that if the rate of twist of the beam were $d\phi/dz$, then the resulting torque would be

$$C\frac{d\phi}{dz} - C_1\frac{d^3\phi}{dz^3}. \tag{4.25}$$

Thus for the case of pure bending of a beam of general cross-section, the governing buckling equation is equation (4.20) if the warping rigidity is neglected. If equation (4.25) is used, however, then the governing equation becomes

$$C_1\frac{d^4\phi}{dz^4} - C\frac{d^2\phi}{dz^2} - \frac{M^2}{EI}\phi = 0. \tag{4.26}$$

The general solution of this equation is

$$\phi = A_1\sin mz + A_2\cos mz + A_3\sinh nz + A_4\cosh nz, \tag{4.27}$$

where

$$\left.\begin{array}{l} m^2 = \left(\alpha^2 + \beta\right)^{\frac{1}{2}} - \alpha, \\ n^2 = \left(\alpha^2 + \beta\right)^{\frac{1}{2}} + \alpha, \end{array}\right\} \tag{4.28}$$

and

$$\alpha = \frac{C}{2C_1}, \quad \beta = \frac{M^2}{EIC_1}. \tag{4.29}$$

The four constants of integration are found from the boundary conditions.

If the ends of the beam cannot rotate about the z-axis (cf. fig. 4.4) but are free to warp; then both ϕ and $d^2\phi/dz^2$ are zero at $z = 0, \ell$. The constants A_2 and A_4 are both zero, and

$$
\left.
\begin{aligned}
& A_1 \sin m\ell + A_3 \sinh n\ell = 0 \\
\text{and} \quad & -m^2 A_1 \sin m\ell + n^2 A_3 \sinh n\ell = 0.
\end{aligned}
\right\} \quad (4.30)
$$

Thus a non-trivial solution can occur only if

$$
(m^2 + n^2)(\sinh n\ell) \sin m\ell = 0, \quad (4.31)
$$

that is, $\sin m\ell = 0$, or $m = \pi/\ell$ for the lowest buckling mode. The critical value of the bending moment is then determined as

$$
M_{cr} = \gamma_1 \frac{\sqrt{EIC}}{\ell}, \quad (4.32)
$$

where

$$
\gamma_1 = \pi \left(1 + \frac{C_1}{C} \frac{\pi^2}{\ell^2} \right)^{\frac{1}{2}}. \quad (4.33)
$$

Equation (4.32) reduces to equation (4.21) if the warping rigidity C_1 is set equal to zero.

Similarly, for Galileo's problem, allowing for warping, equation (4.22) is replaced by

$$
C_1 \frac{d^4\phi}{dz^4} - C \frac{d^2\phi}{dz^2} - \frac{W^2}{EI} (\ell - z)^2 \phi = 0. \quad (4.34)
$$

Timoshenko (1910) solved this equation in terms of an infinite series to give

$$
W_{cr} = \gamma_2 \frac{\sqrt{EIC}}{\ell^2}, \quad (4.35)
$$

and he tabulated values of γ_2 as a function of $\ell^2 C/C_1$. He also showed that, for large values of $\ell^2 C/C_1$, the value of γ_2 was given approximately by

$$
\gamma_2 = \frac{4.013}{\left(1 - \sqrt{C_1/\ell^2 C}\right)^2}, \quad (4.36)
$$

cf. equation (4.24).

These specific solutions of problems of elastic buckling show that, for cases likely to be of practical importance, lateral-torsional buckling must be considered – the analysis must not be confined to the original plane of the member. Moreover, if these elastic studies were to be incorporated in an engineering design process, then account would have to be taken of the limiting yield stress of a real material (it was seen that the empirical approach of equation (4.19), from British Standard 449, introduced an interaction between buckling and yield).

4.6 Horne 1954

As has been seen, the Perry–Robertson approach aims at the calculation of a critical stress σ in terms of the values of yield stress on the one hand and an elastic buckling stress on the other. Equation (4.17), for example, introduces the Euler buckling stress (about the minor axis) σ_E and the yield stress σ_0 into a quadratic equation for σ. The analysis is formulated in terms of stress, but the equations derive from the evaluation of bending moment at a critical cross-section; that moment must be such that buckling of the member as a whole does not occur. The (reasonable) philosophy underlying such an analysis is that a column on the point of becoming unstable has little reserve of strength; once a certain value of bending moment has been reached at a critical cross-section, buckling is imminent. Some remarks are made on this matter in Chapter 10.

In general, the column in a building frame may be acted upon by an axial load in the presence of bending applied at both ends about both axes of the cross-section. Unless bending is confined entirely to the minor axis, failure of the column will occur by lateral-torsional buckling. This problem was investigated in the 1950s by M.R. Horne, who was concerned with the development of the plastic methods for the design of steel frames, and his work is reported extensively in Baker *et al.* (1956).

Horne's basic (I-section) column is illustrated in fig. 4.5, and he distinguishes nine cases of significantly different end conditions, as shown in fig. 4.6. The symbol O_x implies that no bending is applied about the major axis of the section (and similarly for O_y and the minor axis). The symbol E denotes an elastic member connected to the column, which could perhaps provide restraint about the appropriate axis, while P (which derives from 'plastic') implies the application of a fixed and limited bending moment such as would be engendered in the plastic collapse of a frame.

Thus the case $O_x O_y$ is the pin-ended axially loaded 'Euler' column. The case $E_x O_y$ is perhaps the one envisaged by the building codes as of significant importance; the column, part of a building frame, is acted upon by beams supporting loaded floors. The case $P_x P_y$ is discussed extensively by Horne (together with its subclass $P_x O_y$) – the column length is loaded as in fig. 4.5, with known and fixed values of the end couples and a known value of the axial load. A design method, making use of the analyses described earlier in this chapter, is needed to

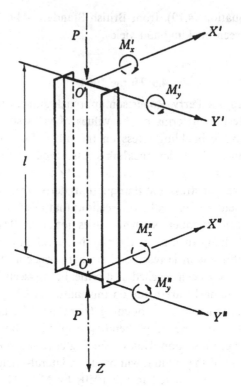

Fig. 4.5. Column length subjected to known axial load and known terminal bending moments (from *The Steel Skeleton* II).

predict (and of course obviate) the onset of buckling. Horne combines ingeniously the results of sophisticated theory (such as that of Prandtl and of Timoshenko) with a practical approach reminiscent of Rankine and Perry.

He starts by considering the case of uniform bending about the major axis $(M'_x = M'_x)$ and no bending about the minor axis $(P_x O_y)$, and quotes a semi-empirical 'safe' interaction formula

$$\left(\frac{M'_x}{M_E}\right)^2 + \left(\frac{P}{P_E}\right) = 1, \qquad (4.37)$$

where P_E is the Euler buckling load (about the *minor* axis), that is $P_E = \pi^2 EI_y/\ell^2$, and M_E is the lateral-torsional buckling moment given by equation (4.21), i.e. $M_E^2 = \pi^2 EI_y C/\ell^2$. The case of non-uniform bending, $M'_x = \beta M'_x$, where $-1 \le \beta \le 1$, is dealt with by defining a moment

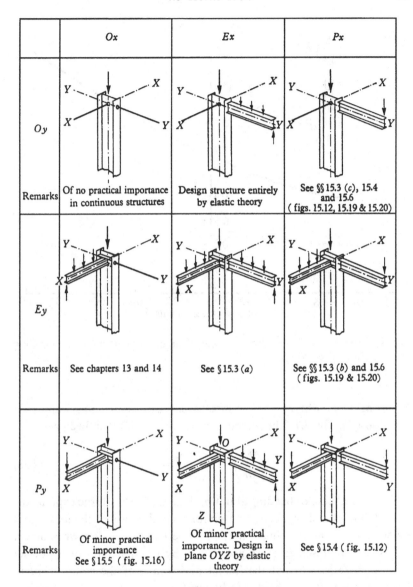

	Ox	Ex	Px
Oy			
Remarks	Of no practical importance in continuous structures	Design structure entirely by elastic theory	See §§ 15.3 (c), 15.4 and 15.6 (figs. 15.12, 15.19 & 15.20)
Ey			
Remarks	See chapters 13 and 14	See § 15.3 (a)	See §§ 15.3 (b) and 15.6 (figs. 15.19 & 15.20)
Py			
Remarks	Of minor practical importance See § 15.5 (fig. 15.16)	Of minor practical importance. Design in plane OYZ by elastic theory	See § 15.4 (fig. 15.12)

Fig. 4.6. Classification of column end conditions (from *The Steel Skeleton* II; paragraph numbers refer to Chapter 15 of that book).

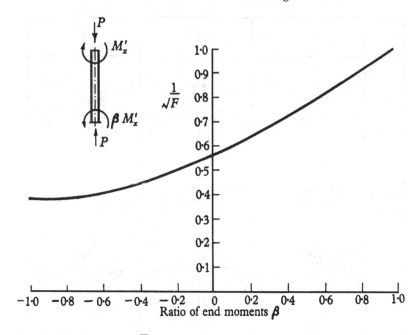

Fig. 4.7. The factor $1/\sqrt{F}$ gives an equivalent uniform moment for buckling analysis (from *The Steel Skeleton* II).

$M_x = M'_x/\sqrt{F}$, where $1/\sqrt{F}$ is most easily expressed numerically as a function of β, fig. 4.7. The interaction formula (4.37) then becomes

$$\left(\frac{M_x}{M_{\mathrm{E}}}\right)^2 + \left(\frac{P}{P_{\mathrm{E}}}\right) = 1. \qquad (4.38)$$

For the full case of bending about both axes $(P_x P_y)$, Horne exploits the simplifying idea of replacing non-uniform bending about the minor axis by similarly using the function $1/\sqrt{F}$. The equivalent uniform moment M_y is then represented (very accurately) by the first term of a Fourier half-range series, so that the lateral deflexion u of the column due to bending (without axial load) can be written

$$u = \left(a + \frac{4}{\pi^3}\frac{M_y \ell^2}{EI_y}\right)\sin\frac{\pi z}{\ell}, \qquad (4.39)$$

where $a \sin \pi z/\ell$ represents the initial (Perry–Robertson) lack of straightness of the column.

If now the column is acted upon by a uniform moment M_x (actually equal to $M'_x/\sqrt{F}$) and an axial load P, and if

$$\left(\frac{M_x}{M_E}\right)^2 + \left(\frac{P}{P_E}\right) = \gamma_y, \qquad (4.40)$$

(cf. equation (4.38)), then it is easily shown that the lateral deflexion of the column increases to

$$u' = \frac{u}{1 - \gamma_y}. \qquad (4.41)$$

When $M_x = 0$, then the equation (4.41) reduces to the familiar expression

$$u' = u\frac{P_E}{(P_E - P)}. \qquad (4.42)$$

From this point the analysis effectively follows that of Perry–Robertson; both the minor-axis bending moment and the major-axis bending moment are calculated at mid-height of the column, and the total stress there related to the yield stress of the material. The whole analysis can be represented by a single chart which can be used for design; conditions at the ends of the column are also checked.

The whole of Horne's analysis, summarized briefly above, can be seen to have developed logically from the elastic theories of the eighteenth century. The work is applied finally to columns in which one end (or both) has developed a plastic hinge (a case which is important, for example, in the design of industrial portal frames). As will have been seen, the work is based firmly on an accurate mathematical description coupled with empirical variables which allow for practical imperfections (which are known only on a statistical basis). Many components of the analysis (for example, torsional stiffnesses and warping factors, Perry–Robertson constants, equivalent factors for non-uniform bending moments) are incorporated in British Standard 5950: Part 1: 1990 (Structural use of steelwork in building).

The design method may be used for any case for which the end conditions for a particular column length can be specified. If a particular section appears to be marginally unsatisfactory, then the substitution of a slightly larger section for the column (very often without penalty of weight or cost) is likely to ensure stable behaviour. However, the exact specification of the conditions for which a column must be designed is not necessarily a simple matter. This question is also addressed further in Chapter 10.

5

The Masonry Arch

As was mentioned in Chapter 2, Coulomb's memoir of 1773 made contributions to each of the four major problems of civil engineering in the eighteenth century – the strength of beams, the strength of columns, the thrust of soil and the thrust of arches. In all four of these topics Coulomb made advances by considering closely the basic equations of equilibrium, both for the structure overall and at imaginary internal cuts. The work on the fracture of beams has been summarized in Chapter 2.

For the next two problems, columns and soil, Coulomb studies failure planes along which slip is occurring, resisted by the cohesion and friction of the material. That is, just as for the beam problem, solutions are obtained from equilibrium equations combined with a knowledge of material properties. These solutions, and their relation to previous work, are described further in Heyman (1972). By contrast, Coulomb's solutions for arches make only marginal reference to the strength of the material (masonry), and his exploration of the stability of the arch is based solely on considerations of equilibrium, coupled (as is made explicit in the title of his memoir) by principles of maximum and minimum. Indeed the arch seems to have been regarded as a problem separate from other studies in the development of structural mechanics, at least until the end of the nineteenth century.

The arch problem was enumerated by Gautier in 1717, in his book on bridge abutments, when he listed 'cinq Difficultez proposées aux Sçavans, à resoudre', namely

1. the thickness of abutment piers for all kinds of bridges;
2. the dimensions of internal piers as a proportion of the span of the arches;

3. the thickness of the voussoirs between intrados and extrados in the neighbourhood of the keystone;
4. the shape of arches; and
5. the dimensions of retaining walls to hold back soil.

Coulomb's fourth problem, the thrust of arches, was directed to the determination of the abutment thrust, so that the abutments themselves could be designed; this is Gautier's problem No. 1, and it is the problem to which the whole of the analysis was directed. In his problems 2, 3 and 4, Gautier recognizes that the solution requires the examination of both the shape and the thickness of the arch.

There is no question here of fracture or of ultimate load of the structure, as there is for Galileo's beam; that is, the *strength* of the arch is not under examination. (In fact, as will be seen, ideas of fracture were introduced quite early in order to clarify the analysis, but these ideas did not stem from failure of the arch itself.) Rather it was the 'working state' that was considered; that arches thrust against their abutments was obvious, and what was needed was a way of calculating numerically the value of that thrust.

5.1 Robert Hooke 1675

Robert Hooke knew, in a physical sense, how arches worked. It was his job as Curator of Experiments to the Royal Society to prepare demonstrations for that body; he held the post from 1663 to his death 40 years later. Among the host of such demonstrations in all fields of science he showed experiments on model arches, but he could not, at the time (or, indeed, at any time), provide the corresponding mathematical analysis. Instead, he published in 1675, in a totally unrelated book (on helioscopes and other instruments), a series of anagrams 'to fill the vacancy of the ensuing page'. Among these, No. 2 yields the famous *ut tensio sic vis*; No. 3 is concerned with 'the true Mathematical and Mechanichal form of all manner of Arches for Building', and, deciphered, reads '*ut pendet continuum flexile, sic stabit contiguum rigidum inversum*'.

As hangs the flexible line, so but inverted will stand the rigid arch. This solution to the anagram was not published until after Hooke's death, but Hooke knew that if he could solve the problem of the shape of the catenary, he would at the same time have found the shape of the perfect arch to carry the same loads in compression. The problem of the catenary was not easy; as has been noted, Leibniz appears to have obtained the solution in 1691, and indeed Huygens and John Bernoulli also seem to have solved the problem at about the same time. In the climate of extreme secrecy and competition in which these scientists worked, however (the same climate that led Hooke to cloak his discoveries in the disguise of virtually indecipherable anagrams), their statements were not fully supported by mathematical proofs. It fell to David Gregory to publish the mathematics in 1697.

Gregory made mistakes in his analysis, but it seems clear from his commentary why Coulomb should have cited the paper some 75 years later. In Ware's (1809) translation (from the Latin), Gregory states:

In a vertical plane, but in an inverted situation, the chain will preserve its figure without falling, and therefore will constitute a very thin arch, or fornix; that is, infinitely small rigid and polished spheres disposed in an inverted arch of a cateneria will form an arch; no part of which will be thrust outwards or inwards by other parts, but, the lowest part remaining firm, it will support itself by means of its figure ... And, on the contrary, none but the catenaria is the figure of a true legitimate arch, or fornix. *And when an arch of any other figure is supported, it is because in its thickness some catenaria is included.* Neither would it be sustained if it were very thin, and composed of slippery parts. From Corol. 5 it may be collected, by what force an arch, or buttress, presses a wall outwardly, to which it is applied; for this is the same with that part of the force sustaining the chain, which draws according to a horizontal direction. For the force, which in the chain draws inwards, in an arch equal to the chain drives outwards.

Here, then, is Gregory's complete grasp of the end to which the analysis of the arch is directed; the horizontal component of the abutment thrust of an arch has the same value as the horizontal pull exerted by the equivalent hanging chain. Further, the italicized statement (italics added by Ware) is extremely powerful; in modern terms, Gregory asserts that if any thrust line can be found lying within the masonry, then the arch will stand.

5.2 La Hire 1695, 1712

Gregory's approach to the arch problem, by finding a solution to the analogous problem of the hanging chain, was, as will be seen, explored further some 50 years later by Poleni in his study of the dome of St Peter's, Rome. Immediately, however, a direct solution was attempted – La Hire (1695) considered the statics of a semi-circular arch assembled from stone wedges (voussoirs). The difficulty lay in the assumptions to be made about the behaviour of the voussoirs; La Hire assumed that the joints between the stones were frictionless. He then set himself the problem of finding the weights of the voussoirs (for the arch of known semi-circular geometry) so that equilibrium should be maintained. The solution lay in La Hire's invention of the force polygon and the corresponding funicular polygon for the arch (which is nothing more than the line of thrust, or the shape of the inverted hanging chain, although La Hire did not use these terms). For an arch with smooth voussoirs the line of thrust must be perpendicular to the joints, so that the funicular polygon is fixed by the shape of the arch; working backwards, the force polygon can be constructed, and finally the weights of the voussoirs found.

If the springing lines of the arch are horizontal it follows that the weights of the springing voussoirs must be infinite; a finite arch of this form with smooth voussoirs cannot stand. La Hire reached this conclusion as a result of an unrealistic assumption about the behaviour of the material, and in fact he noted that in practice friction between the voussoirs would confer the necessary stability. However, he left the matter there for the time being; the solution to the arch problem was not much advanced, although some valuable tools had been invented.

La Hire returned to the arch in 1712, and he abandoned the assumption of smooth voussoirs; instead, friction was taken to be so large that sliding could not occur. Thus the direction of the line of thrust within the arch was no longer fixed as being perpendicular to the joints, and there was no such simple starting point for the statical analysis. La Hire introduced, apparently for the first time, a clarifying idea of great power. The object of the analysis was to determine the value of the arch thrust, so that the abutments could be designed; if those abutments were too weak, and gave way slightly, how would the arch behave? La Hire stated that the arch would break at a section somewhere between the springing and the keystone.

In fig. 5.1 the joint LM is taken to be critical, and, at that joint, the (slightly) increased span is accommodated by a 'hinge' developing between the voussoirs at L. Thus contact between the upper portion LMF of the arch and the lower portion LMI is only at the point L, and it is through this point that the forces within the arch must pass. This concept unlocks the statics of the arch; in fig. 5.2, the thrust P at the hinge point L must act tangentially to the intrados, and, knowing the weight of the upper portion LMF of the arch (fig. 5.1), the value of all the forces may be found. Finally, by taking moments about H for the lower portion of the arch and the pier, fig. 5.2, an expression to check the stability of the whole structure can be obtained.

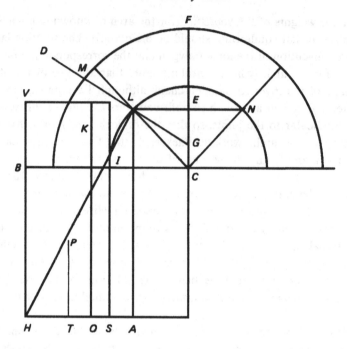

Fig. 5.1. The mechanics of the semi-circular arch (after La Hire (1712)).

La Hire gave no rule for finding the critical point L. Moreover, the presence of two hinges, L and N in fig. 5.1, does not (as will be seen) imply that the arch is necessarily in a critical state. However, La Hire's analysis is a major contribution to the problem, and the resulting estimate of the abutment thrust is greater than the minimum necessary for stability, so that the calculation is, as it happens, 'safe'.

As was noted in Chapter 2, Bélidor's section on arches in his *Science des ingénieurs* (1729) is based firmly on this work of La Hire. There are some differences – the weakest section of the arch, LM in fig. 5.1, is taken to be at 45° (i.e. half-way between the springing and the crown), and the thrust acts not at L but at the midpoint of LM. Thus the abutment thrust has value $\sqrt{2}W$, where W is the weight of the 'voussoir' LMF, and this value is again 'safe'. Bélidor's intention, here as elsewhere in his manual of civil engineering, was not necessarily to advance theory, but to establish sets of engineering design rules based upon existing scientific work.

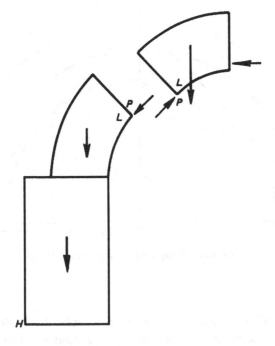

Fig. 5.2. The statics of the arch (after La Hire (1712)).

5.3 Couplet 1729, 1730

It was against this background that Couplet wrote his two remarkable memoirs on arch thrust, in 1729 and 1730. The first of these papers effectively repeats La Hire's analysis of the frictionless case, and Couplet was aware that the work was of little practical application. However, he did make an interesting calculation on the forces imposed by an arch on its centering during construction (Heyman (1976)); this important practical problem had been tackled by Pitot in 1726.

Couplet made a major advance in his second memoir. In his introduction he states precise assumptions about material behaviour – he notes that friction in practice locks the voussoirs together against sliding, while no resistance is offered to separation of the voussoirs. He does not remark on the strength of the stone of which the voussoirs are made, and by implication he assumes that ambient stresses are so low that crushing strength is of little importance.

Thus Couplet makes in effect three key postulates about the behaviour of masonry – it has no tensile strength, infinite compressive strength,

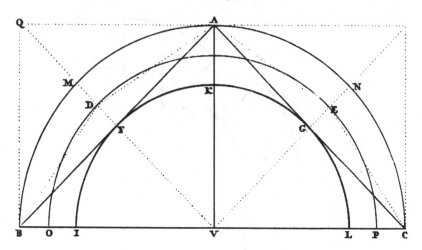

Fig. 5.3. An arch that cannot collapse under a vertical load at the crown (from Couplet (1730)).

and sliding failure cannot occur. Further, Couplet demonstrates in his work the two ways of approaching any structural problem – through equilibrium (statics), in which thrust lines are considered, and through deformation (mechanisms) in which patterns of hinges are constructed.

His proof of his first theorem in the 1730 memoir contains precisely these dual aspects of structural analysis. The theorem states that an arch will not collapse if the chord of half the extrados does not cut the intrados, but lies within the thickness of the arch. Couplet has in mind a semi-circular arch of negligible self-weight subjected to a single vertical point load at the crown A, fig. 5.3. Whatever the magnitude of the load, supporting forces can be generated directly from the abutments B and C, following the straight thrust lines AFB and AGC. Further, says Couplet, for the arch to collapse the angle BAC must open, and this can occur only as a consequence of spread of the abutments (which is ruled out); there is in fact no arrangement of hinges in the extrados and intrados of the arch sketched in fig. 5.3 that is both compatible with thrust lines for the load and also gives rise to a mechanism of collapse.

The next problem tackled by Couplet is to find the least thickness of a semi-circular arch, carrying only its own weight. The arch, says Couplet, will collapse by separating into four pieces, attached to each other by hinges R, T, A, K and F, fig. 5.4. The hinges T and K at the haunches are placed at 45°; Couplet considers the equilibrium of the arch in this

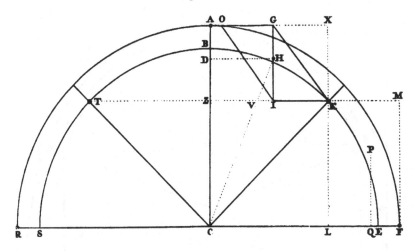

Fig. 5.4. Semi-circular arch of minimum thickness collapsing under its own weight by the formation of hinges (from Couplet (1730)).

state, and finds a single expression relating the thickness of the arch to its (mean) radius R, namely $t/R = 0.101$. The statics are evident in fig. 5.4. For the equilibrium of the piece AK of the arch, the horizontal thrust at A combined with the weight acting through H leads to a thrust at K in the line GK. Now GK is not tangential to the intrados at K, implying that, impossibly, the thrust should escape from the masonry. Couplet misses this point, but the work is otherwise correct. (The intrados hinges actually form at 31° from the springing rather than at 45°, but the analysis is not sensitive to their exact position, and the correct value of t/R is increased only to 0.106.)

Later, in order to determine the value of the abutment thrust for a more general shape of arch, Couplet abandons this kind of collapse analysis, and reworks the La Hire/Bélidor approach; forces are referred to the centre line SX of the arch, fig. 5.5. The thrust at the crown acts horizontally at S, and the weight of half the arch in the line LR; a simple triangulation of forces then gives the magnitude of the abutment thrust, acting in the line LX. Finally, the dimensions of the piers can be calculated so that the whole construction is stable.

Couplet's contribution is outstanding. He had clear ideas of lines of thrust, and of mechanisms of collapse caused by the formation of hinges; he made explicit his simplifying assumptions; and he used these ideas to obtain an essentially correct and complete solution to the problem of

The Masonry Arch

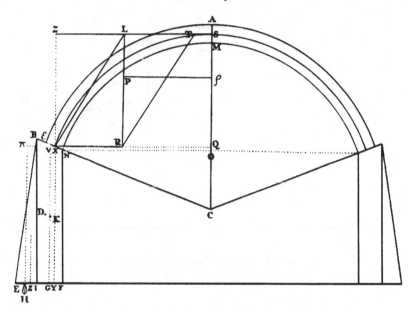

Fig. 5.5. Couplet's force system for design of abutments (from Couplet (1730)).

arch design. His work was noted immediately, and found its way into standard texts (for example, that of Frézier (1737–39)).

In 1732 Danyzy obtained experimental confirmation of the correctness of Couplet's approach. The work was done in Montpellier, and published obscurely and not until 1778. However, Frézier has a plate based upon one of Danyzy's illustrations, showing the collapse of arches made from plaster voussoirs, fig. 5.6. (Figure 241 is half of Couplet's arch, fig. 5.3.) All the arches shown are on the point of collapse, the piers having minimum dimensions. Figure 235, for example, corresponds to the collapse mechanism predicted by Couplet, fig. 5.4. The flat arch of Figure 240 is, within the framework of the assumptions, infinitely strong; it is collapsing only because the abutments are tilting.

5.4 The dome of St Peter's

By about 1740, then, the mechanics of the arch was well understood – not only had this structural problem been solved, but the theory could be applied more generally to the analysis of masonry. Poleni was appointed in 1743 to report on (Michelangelo's) dome of St Peter's, Rome,

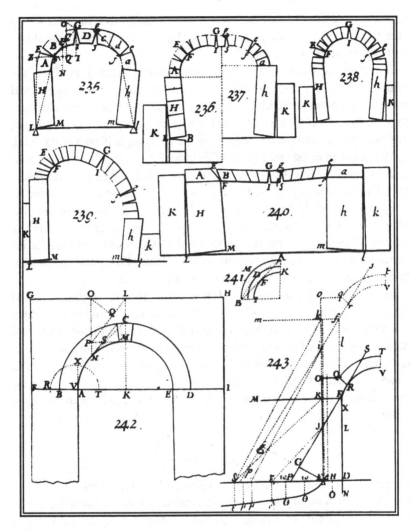

Fig. 5.6. Danyzy's experiments on model arches (from Frézier (1737–39)).

constructed two centuries earlier, and his *Memorie istoriche*, published in 1748, gives a comprehensive review of the existing state of knowledge. He knew the work of La Hire and of Couplet; from fig. 5.7 it will be seen (fig. XII) that he knew of Hooke's hanging chain. He also quoted Gregory, and an interesting development of Gregory's work by Stirling (1717). (Fig. XI, an inverted catenary formed by the balancing of smooth spheres, is based closely upon an illustration by Stirling.)

Echoing Gregory, §5.1 above, Poleni states explicitly that in order for an arch to be stable, all that is necessary is that the line of thrust should lie everywhere within the masonry.

The dome of St Peter's had developed cracks running up from the drum along generators, and dying out near the crown; these meridional cracks divided the dome into portions approximating half-spherical lunes (orange slices). Poleni sliced the dome hypothetically into 50 such lunes, one of which is shown schematically as the tapering half arch of fig. XIII, fig. 5.7; he then considered the equilibrium of this quasi two-dimensional arch tapering to zero thickness at the crown. The thrust line was determined experimentally by loading a flexible string with a series of unequal weights, each weight being proportional to that of a segment of the lune (due allowance was made for the weight of the lantern surmounting the eye of the dome). Figure 5.8 shows the experimental result; the inverted chain lies within the thickness of the dome. The sliced arches were therefore safe, and so, *a fortiori*, was the whole cracked dome.

In 1743 Le Seur, Jacquier and Boscovich, the *tre mattematici*, had made an alternative study of the dome and had concluded that extra ties were needed at the base to contain the horizontal thrust. They had made an estimate of this thrust by an early application of virtual work (a technique which has become of increasing importance, as will be seen in Chapters 7 and 9). Poleni agreed that extra ties should be installed; his own estimate of tie force came from the horizontal pull of the hanging chain.

5.5 Coulomb 1773

Coulomb provided the final pieces of mathematical rigour to complete the theory of masonry arches, and he did this without knowing of the work of Couplet or of Poleni, although La Hire's analysis had found a place in Bélidor's handbook, as has been seen. However, Coulomb as a young man had lived in Montpellier, and he knew Danyzy there; it seems certain that he knew of the collapse of arches by the formation of hinges. Indeed, he concludes that in practice failure will always occur by hinging between voussoirs, and he makes the assumption of slip being prevented by friction. His work on the fracture of columns makes him aware of the possibility of crushing at a hinge point, and he shows how to allow, if necessary, for a finite area involved in transmitting a thrust (in order to avoid the theoretically infinite stresses induced by line contact).

Figure 5.9 reproduces one of Coulomb's figures of 1773; the half arch

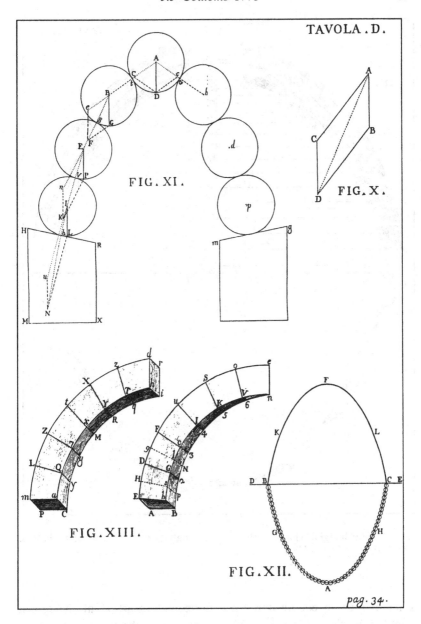

Fig. 5.7. Illustrations of the mechanics of masonry arches (from Poleni (1748)).

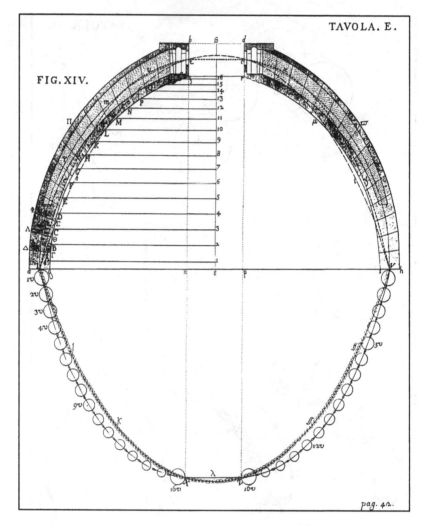

Fig. 5.8. The hanging chain applied to the analysis of the dome of St Peter's, Rome (from Poleni (1748)).

is maintained in equilibrium by a horizontal thrust *H* at the crown (supplied by the other half arch) acting through the point f. Failure is occurring at the joint Mm between voissoirs. Coulomb considers first that hinging occurs about point M in the intrados, and he writes a general expression for the value of *H*. He then shows that the position of M must be chosen so that the value of *H* is maximized; the hinge cannot

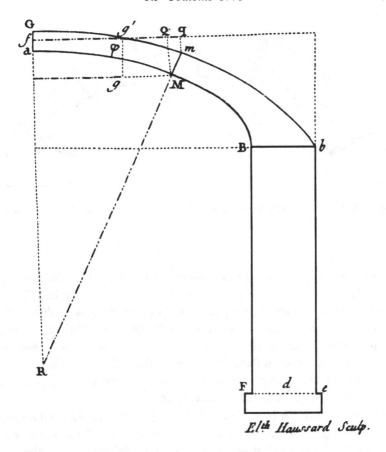

Fig. 5.9. The equilibrium of the arch, as considered by Coulomb (1773).

be placed arbitrarily (as, for example, at 45°). Moreover, this maximum value of *H*, once found, is the *minimum* value for which the arch will be stable. Similarly, if failure is occurring by hinging about the point m in the extrados, then m must be chosen so that the value of H is minimized, and this is the *maximum* value for which the arch will be stable.

This seems to be the first statement that *bounds* could exist on the value of a structural quantity. The idea was demonstrated physically at the Institution of Civil Engineers by Barlow (1846), who started his paper by acknowledging the work of Coulomb. He knew of the equivalence of the line of thrust and the hanging chain, and he showed models of arches of minimum thickness for stability. In another experiment six voussoirs were assembled as in fig. 5.10, with the 'mortar' in each joint in the

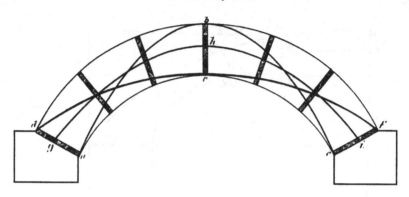

Fig. 5.10. Barlow's model voussoir arch (1846) demonstrating alternative positions of the thrust line.

form of four small pieces of wood, each of which could be withdrawn by
hand. Three out of the four pieces were then indeed removed, in different
configurations, and alternative positions of the thrust line were thus made
'visible'; three positions were sketched by Barlow in his illustration. The
steepest curve, touching the crown at the extrados, is called by Barlow
'the line of resistance', and the flattest curve 'the line of impression'. They
represent the limits corresponding to the least and greatest values of the
horizontal component of the abutment thrust.

Coulomb's work was assimilated slowly into the technical education
of French engineers. The *École des Ponts et Chaussées*, for example, was
well aware of his contributions, and the second edition (1833) of Navier's
Léçons devotes about 50 pages to arch theory. (Prony had taken forward
Coulomb's work on soil mechanics in 1802.) Development of graphic
statics in the nineteenth century, which made lines of thrust part of the
designer's stock in trade, took place after the masonry arch had become
obsolescent; the 1830s saw the construction of the last of the large span
masonry bridges.

A definitive exposition of arch theory was given in 1845 (published
1854) in a long memoir by Yvon Villarceau. He knew that the arch
was essentially statically indeterminate, and that there existed therefore
an infinite number of equilibrium states; he developed a 'safe' design
method by requiring the centre line of the arch to coincide with one
of the possible lines of thrust for the given loading. This inverse design
method requires the numerical solution of the equations, and the results
are presented in the form of tables which can be used immediately in

standard calculations by the bridge designer. As a design method there is really nothing to be added.

5.6 The plastic theorems

As will be noted in Chapter 7, Castigliano (1879) applied his elastic energy theorems to, among other practical examples, the masonry bridge; he obtained solutions that made use of the elastic properties of the stone and mortar, and allowed for cracking should the thrust line fall outside the 'middle-third' of the cross-section. Similarly Pippard *et al.* (1936, 1938) made careful tests in the 1930s of model arches with steel voussoirs, and showed that the slightest imperfection of fit (e.g. at the abutments) converted an apparently hyperstatic into a statically determinate three-pin arch. However, Pippard attempted to interpret his results by principles of minimum elastic energy. The justification for the 'equilibrium' approach of Poleni, Coulomb, Barlow and Yvon Villarceau is provided by the twentieth-century plastic theorems (to be discussed more fully in Chapter 9).

The necessary assumptions for the application of plastic theory to masonry are that

     (i)   masonry has no tensile strength,

    (ii)   masonry has infinite compressive strength,

and (iii)   sliding does not occur.

These are, of course, the assumptions made from Couplet onwards; Coulomb discussed necessary modifications if the stone were in danger of crushing. On the basis of these assumptions Kooharian (1953) showed that the analysis of masonry could be interpreted within the framework of plasticity (see e.g. Heyman (1995)).

In fig. 5.11(*a*) a hinge is opening in a voussoir construction; if the axial load being transmitted at the hinge is N, then the effective moment has value $|M| = hN$. The lines $M = \pm hN$ define a permissible region, fig. 5.11(*b*), which is in fact a yield surface of plastic theory. A point in the M, N diagram represents a state of the cross-section being considered. If $|M| < hN$, the eccentricity of the thrust is less than h, and no hinge is formed; for $|M| = hN$ a hinge forms in the extrados or intrados; and $|M| \not> hN$ as thrusts must lie within the masonry.

If the stone of infinite strength is replaced by the real stone with a finite crushing strength, then the yield surface AOB is replaced by the curved boundary $OCDEO$, formed by two parabolic arcs. However, only a small

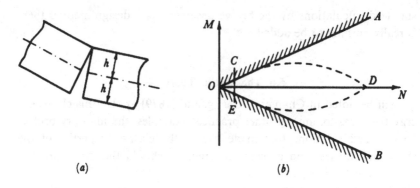

(a)   (b)

Fig. 5.11. The yield surface for masonry.

portion of this real yield surface applies to practical masonry structures. A typical value of permitted stress used in nineteenth-century design of large bridges is 10 per cent of the crushing strength (Yvon Villarceau); nominal stresses are likely to be less than this, but even at 10 per cent the portion of the yield surface is the slightly curvilinear triangle OCE.

Whether or not the straight or slightly curved boundaries are used, the 'safe' theorem of plasticity states that if all stress resultants lie within the yield surface (i.e. within the triangle OCE), then the construction is indeed safe, and cannot collapse. The power of this theorem lies in the fact that it is sufficient to find any one such safe state. In terms of arch construction, if any one of the infinitely many lines of thrust equilibrating the applied loads can be shown to lie within the arch profile, then this is proof that the arch cannot collapse under those loads. This theorem was, as has been seen, stated explicitly by Poleni in his examination of the dome of St Peter's.

Thus design rules for low-stressed masonry construction should be directed to the goal of ensuring that the shape of the structure conforms to the shape required by statics – the two surfaces of the arch must be able to contain the inverted hanging chain. These are rules of geometry, effective for materials working at low stresses (stone, timber), but ineffective, as Galileo saw, for structures that work their materials harder.

6

Elastic Beams and Frames

It was clear to Galileo that a beam resting on three supports (which, in modern terminology, would be hyperstatic) could be subjected to forces not envisaged by the engineer. That is, an accidental imperfection (and Galileo used the word *accidente*), such as decay of one of the end supports, could lead to a set of forces that would break the beam. He was equally clear that no such accident could happen to a beam on two supports; if the supports sink then the beam follows – the statics of a statically determinate beam are unique.

However, it does not seem that Galileo was concerned with any concepts that might stem from the consideration of what is now known as the hyperstatic structure. His objective was, as has been described, to calculate the breaking strength of beams, and for this purpose he determined the greatest value of bending moment in a beam, whether that beam were simply supported or a simple cantilever. The value of bending moment having been found, the problem then became one of the strength of materials, and the historical notes given in Chapter 2 are concerned with the correct way of calculating the moment of resistance of a cross-section.

6.1 Girard 1798

It was noted in passing in Chapter 2 that Mariotte made tests on fixed-ended beams, and that he concluded from his experiments that their strengths were twice those of corresponding simply supported beams. Mariotte gave no theoretical explanation for this result, and the problem of the strength of the hyperstatic beam seems to have remained unexplored throughout the eighteenth century. However, in 1798 Girard (in his book to which reference was made in Chapter 2) gave the analysis for

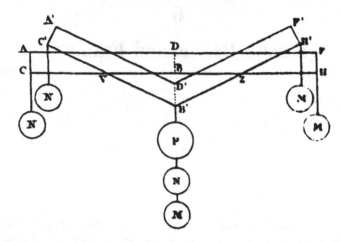

Fig. 6.1. Girard's diagram for the analysis of the breaking strength of a fixed-ended beam.

the breaking strength of a fixed-ended beam and of a propped cantilever; in fig. 6.1 is reproduced his diagram, showing a 'virtual' displacement he used in his calculations.

Figure 6.2 shows the beam sketched in a more modern convention; it is simply supported at the points V and Z. If the beam has moment of resistance B at the loading point, then in the case $M = N = 0$ the collapse load has value

$$P = B\frac{(g+f)}{gf}. \qquad (6.1)$$

If, however, the loads M and N have positive values, then the value of the collapse load P will be increased, and Girard uses the sketched displacement of fig. 6.1 to solve this problem. He then considers the case in which the value of N is just sufficient to cause fracture at the support V, where the moment of resistance has value B', and similarly for the load M, and he shows that the total load that can be sustained is

$$P + P' = B\frac{(g+f)}{gf} + \frac{B'g + B'f}{gf}. \qquad (6.2)$$

Further, for a prismatic beam $B' = B' = B$, and

$$P + P' = 2B\frac{(g+f)}{gf}, \qquad (6.3)$$

which completes the proof that the strength of a fixed-ended beam, with ends restrained at V and Z, is twice that of the simply supported beam.

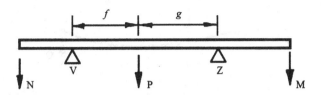

Fig. 6.2. Figure 6.1 redrawn.

Finally, Girard shows that the strength of a uniform cantilever (fixed at V and simply supported at Z) is

$$P + P' = B\frac{(2g + f)}{gf}. \tag{6.4}$$

6.2 Navier 1826

Girard's results, then, are those of a collapse analysis; the idea of an ultimate moment of resistance is accepted, and an examination of the structure then follows. Girard extends his method briefly to a discussion of a continuous beam resting on a number of equally spaced supports. It fell to Navier, some 25 years later, to examine the problem of the continuous beam from the point of view of elastic behaviour.

As was mentioned in Chapter 4, it seems that Navier was the first to formulate a general small-deflexion bending theory by taking one of the co-ordinate axes to lie along the initial direction of the (straight) member. The curvature under the action of a bending moment M can then be approximated by d^2y/dx^2, leading to the basic elastic equation

$$EI\frac{d^2y}{dx^2} = M. \tag{6.5}$$

If the bending moment M is a function of x only, this equation is straightforward, and it was seen that Daniel Bernoulli (1741) had in fact written the equation for a cantilever, equation (4.4), and integrated in quadratures to obtain an expression for the deflected shape. Similarly Euler (1757) had solved equation (6.5) for the case $M = -Py$, and obtained directly 'Euler' buckling loads.

Navier in his *Leçons* of 1826 assimilates all this information in his notes for the students of the *Ponts et Chaussées*. The second volume of this text discusses problems in fluids and in design of machine elements.

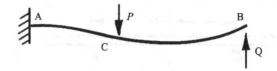

Fig. 6.3. Propped cantilever with a single redundancy.

The first volume is divided into four sections, the first of which is concerned with 'strength of materials'; it was this section alone that Saint-Venant expanded at such length, and that was discussed in Chapter 2 in connexion with Galileo's problem. The second section of Navier's first volume deals with geotechnical problems, and the third with the behaviour and design of masonry arches.

The fourth section is concerned with the behaviour and design of timber structures. Beams in 1826 were, of course, in general made from wood, and it is this section of Navier's *Leçons* which deals with bending. After establishing equation (6.5), Navier tackles problems of hyperstatic indeterminacy in connexion with trusses, and this work will be referred to again in Chapter 7. For his first bending problem, that of a propped cantilever, fig. 6.3, Navier introduces a single redundant quantity (an unknown reaction Q at the simple prop), so that the bending equation, the second-order differential equation (6.5), contains one unknown. A double integration introduces two further unknowns, and three boundary conditions (zero deflexion at both ends and zero slope at the fixed end) suffice to evaluate the three unknown quantities and to solve the problem. Navier's notation is very close to that used today, and his methodology for finding the elastic solution remains unchanged, although different techniques of calculation were soon developed.

For example, Navier wrote separate differential equations for the portions AC and CB of the beam in fig. 6.3, and matched solutions by ensuring that they gave the same slope and deflexion of the beam at the common point C. Thus in fact four constants of integration are involved, and it was not long before techniques were introduced to deal economically with discontinuities such as are engendered by the point load P in fig. 6.3.

The beam sketched in fig. 6.4 is acted upon by a series of point loads, which divide the length into sections 1, 2, ... The differential equations

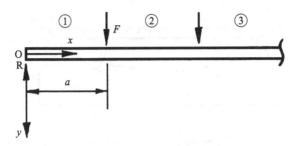

Fig. 6.4. Long beam acted upon by several loads.

of bending for the first two sections may be written

$$\left. \begin{array}{c} EI\dfrac{d^2y_1}{dx^2} = -Rx, \\[2mm] \text{and} \quad EI\dfrac{d^2y_2}{dx^2} = -Rx + F(x-a), \end{array} \right\} \tag{6.6}$$

and integration of these equations will give rise to four arbitrary constants, two of which may be found by expressing continuity of slope and deflexion at $x = a$. However, Clebsch (1862) showed that the equations could be written with only two constants, independently of the number of sections of the beam being considered, if they were integrated in the form

$$\left. \begin{array}{c} EIy_1 = -\dfrac{Rx^3}{6} \qquad\qquad\quad +\alpha x + \beta, \\[2mm] EIy_2 = -\dfrac{Rx^3}{6} + \dfrac{F}{6}(x-a)^3 \;\; +\alpha x + \beta. \end{array} \right\} \tag{6.7}$$

Evidently at $x = a$ (and at $x = b$ and so on) the required continuity is achieved. It is also evident that a (hyperstatic) beam resting on several supports can be analysed in the same way.

6.3 Slope-deflexion equations

Navier made explicit the elastic basis of calculation of such hyperstatic structures. The equations of equilibrium are insufficient to determine the required structural quantities, and have to be afforced both by the (elastic) laws of deformation of the members, e.g. equation (6.5), and by the conditions of compatibility (members must be joined to each other in a prescribed way, and the structure has to satisfy specified

Elastic Beams and Frames

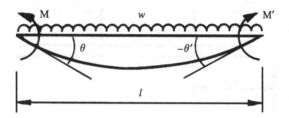

Fig. 6.5. Notation for slope-deflexion equations.

support conditions). Deeper examination of the hyperstatic structure – the idea that such a structure is capable of sustaining states of self-stress – seems not to have been undertaken by Navier. However, the principles were clear, and, as with Clebsch noted above, attention continued to be directed to reducing the labour of calculation.

It was appreciated that the equations to be solved were linear in the force quantities (loads and reactions, and bending moments); the problem was how to simplify the solution of a large number of linear simultaneous equations. Clapeyron (1857) was the first to write elementary 'slope-deflexion' equations. In fig. 6.5 a uniform beam is acted upon by a uniformly distributed load w and end couples M and M'; then the end slopes (clockwise positive) may be written

$$\left. \begin{aligned} 6EI\theta &= \tfrac{1}{4}w\ell^3 - \left(2M + M'\right)\ell, \\ \text{and} \quad -6EI\theta' &= \tfrac{1}{4}w\ell^3 - \left(M + 2M'\right)\ell. \end{aligned} \right\} \tag{6.8}$$

Clapeyron then considers a continuous beam of n spans, so that $2n$ equations may be written involving, linearly, $4n$ quantities like θ, θ', M and M'. At each support the moments and slopes are equal, so that a further $2(n-1)$ equations may be written, and there are thus $(4n-2)$ equations for $4n$ unknowns. If, for example, the two ends of the continuous beam are simply supported, then two values of the bending moment are known to be zero, and the problem may be solved; alternatively, a clamped end will introduce a known zero slope, and again the number of equations is sufficient.

Bertot (1855) is sometimes credited with 'Clapeyron's' theorem of three moments, since (acknowledging Clapeyron) he applied equations (6.8) to a continuous beam, simply supported at its ends, and showed how numerical solutions could be easily found. If the slopes given by equations (6.8) are equated at each support in turn, then

$$\left.\begin{aligned}
2(\ell_1 + \ell_2)M_1 + \quad\quad \ell_2 M_2 \quad\quad &= \tfrac{1}{4}\left(w_1\ell_1^3 + w_2\ell_2^3\right) \\
\ell_2 M_1 + 2(\ell_2 + \ell_3)M_2 + \ell_3 M_3 &= \tfrac{1}{4}\left(w_2\ell_2^3 + w_3\ell_3^3\right)
\end{aligned}\right\} \quad (6.9)$$

and so on.

If now some value is assumed for M_1, the first of equations (6.9) may be used to calculate M_2, the second will give M_3, and so on, until finally the last equation will give a value of M_n, which should in fact be zero. A second assumed value for M_1 will give a different value for M_n; since the problem is linear, interpolation will give the correct value for M_1 for which M_n is zero, and hence correct values for all the other unknowns.

6.4 Hardy Cross 1930

Such numerical techniques were pursued in various guises for the best part of the following century – elastic theory applied to beams and frames generated large numbers of linear simultaneous equations, which had to be solved by hand calculation. For example, more than 50 years after Clebsch (1862), the 'discontinuity' notation was tidied up by Macaulay (1919), was generalized by Wittrick in 1965 to include problems involving axial load, and was embraced by Lowe in 1971 in his *Classical theory of structures*.

Pursuing a different path, Hardy Cross developed in 1930 his method of moment distribution, in which a partial solution for a modified frame is altered systematically to lead to the correct solution. Although the operation of the method involves, apparently, some (imaginary) adjustments to the actual structure, it is in reality yet one more way of obtaining approximate solutions to a set of linear equations; the approximation may be made as exact as is wished by carrying the process on for a sufficient number of stages.

Some preliminary lemmas are needed, of which two are illustrated in fig. 6.6; these lemmas follow directly from equations (6.8). In fig. 6.6(a) a number of members, whose remote ends are clamped, are all connected together rigidly at a single node. The stiffness of each member in bending is defined as $k = EI/\ell$, where EI is the flexural rigidity in bending and ℓ the length of the member. If, then, a couple M is applied to the node causing the ends of all the members meeting there to turn through the same angle, then an end couple will be induced in member r of magnitude $M_r = (k_r/\Sigma k_r)M$; the bracketed term is the 'distribution factor' of

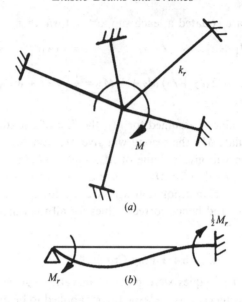

Fig. 6.6. Structural elements for the moment-distribution process.

the member. A second application of equations (6.8) is illustrated in fig. 6.6(b). If a moment M_r is applied to the end of member r, then a moment $\frac{1}{2}M_r$ is induced at the far, clamped, end; the 'carry-over factor' is $\frac{1}{2}$. (More complex expressions for distribution and carry-over factors may be written for members carrying axial load, and for members having their remote ends pinned rather than fixed.)

The example of a three-span uniform beam, fig. 6.7(a), will make the method clear. The first stage in the process is to apply external 'clamps' to each joint, so that the continuous beam is replaced by three separate fixed-ended sections. The external loads are then applied – in this example, the single point load will induce 'fixed-end moments' of value 300 (arbitrary units) as shown in fig. 6.7(b). The left-hand fixed end can supply the required fixed-end moment of 300, but evidently the moment of 300 acting on the clamp at B must be supplied by some external agency. The clamp at B is now 'released', that is, an anticlockwise moment of 300 is applied at B, fig. 6.7(c). By the two lemmas of fig. 6.6, the moment of 300 splits as 150/150 at B since the two members meeting there have equal stiffnesses, and moments of 75 are induced at both A and C. Figures 6.7(b) and (c) are now superimposed to give the state of fig. 6.7(d), in which joint B is now 'in balance', but joint C requires

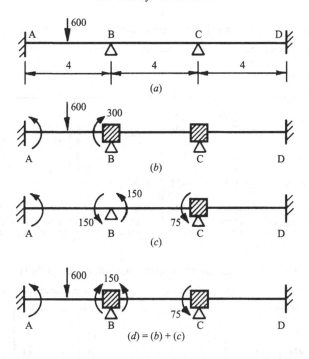

Fig. 6.7. The first stage of a moment-distribution process. (a) The loading, (b) all joints clamped, (c) external couple applied at B and (d) joints reclamped.

the application of an external couple of magnitude 75. Joint B is now 'reclamped', and joint C is 'released' (i.e. a clockwise couple of 75 is applied), inducing 'carry-over' moments at B and D.

The whole process can be tabulated easily, as shown in Table 6.1. The process has been abandoned with joint C slightly out of balance, but it will be noted that convergence is rapid. The computational method can be extended to complex frames consisting of arrays of beams and columns, and full allowance can be made for lateral deflexions (sway) of such frames.

Hidden in the example of fig. 6.7 is a fundamental distinction between the force and displacement methods of elastic structural analysis. In the conventional view of a hyperstatic structure, the three-span beam with fixed ends has four redundant force quantities, whose values are to be determined. There is some choice as to how these four quantities are assigned physically, but they could be the four bending moments in the beam at the fixed ends *A* and *D* and at the internal joints *B* and *C*.

Table 6.1. *Moment distribution for the beam of fig. 6.7*

	A		B		C		D
Distribution factors		$\frac{1}{2}$	$\frac{1}{2}$		$\frac{1}{2}$	$\frac{1}{2}$	
Fixed-end moments	**−300**	**300**					
Balance B	−75	−150	−150	−75			
Balance C			18.8	37.5	37.5	18.8	
Balance B	−4.7	−9.4	−9.4	−4.7			
Balance C			1.2	2.3	2.4	1.2	
Balance B	−0.3	−0.6	0.6	−0.3			
Stop.							
Totals:	−380	140	−140	−40.2	39.9	20	
Exact values	−380	140	−140	−40	40	20	

Table 6.2. *The problem of fig. 6.7 solved in symbols*

	A		B		C		D
Apply x at B	**−300**	**300**					
	$\frac{1}{4}x$	$\frac{1}{2}x$	$\frac{1}{2}x$	$\frac{1}{4}x$			
and y at C			$\frac{1}{4}y$	$\frac{1}{2}y$	$\frac{1}{2}y$	$\frac{1}{4}y$	

Indeed, it is the values of these four bending moments that are displayed in the last line of Table 6.1.

However, the Hardy Cross method works in fact with displacement rather than force variables; although it is not immediately apparent, the beam of fig. 6.7 is regarded as having only two unknown quantities whose values are to be determined, and these are the rotations of the beam at the internal supports B and C (whose values are not actually calculated). Thus a system of four simultaneous equations is replaced by only two equations in Hardy Cross's method, as will be evident if the distribution process of Table 6.1 is carried out with symbols, as in Table 6.2.

If the values of x and y in Table 6.2 are such that 'balance' has been achieved exactly, then

$$\left. \begin{array}{ll} \text{at } B & 300 + x + \frac{1}{4}y = 0 \\ \text{and at } C & \frac{1}{4}x + y = 0, \end{array} \right\} \quad (6.10)$$

from which $x = -320$ and $y = 80$, giving the last line of Table 6.1.

The starting quantities for the moment distribution process are the 'fixed-end moments'. The first line of Table 6.1 gives these moments for a single central point load on span *AB*, fig. 6.7, and it is clear that any other loading on the beam can be dealt with span by span for the clamped configuration of fig. 6.7(*b*) – all that is needed is a straightforward calculation for a single-span clamped beam in order to give the required fixed-end moments. Bendixen (1914) had introduced fixed-end moments in his development of slope-deflexion equations for the solution of continuous beams and frames; these equations are written in terms of displacement variables (rotations at the joints), but he did not anticipate Hardy Cross in proposing a systematic method of approximate solution.

In retrospect, Hardy Cross's method could be classified as a 'relaxation' technique. Southwell (1940) developed his relaxation methods for dealing, in general, with (two-dimensional) continuum problems; the governing partial differential equation for a given problem is replaced by finite-difference equations whose solutions are sought at a large number of discrete nodal points embracing the field under study. These finite-difference equations are, in effect, linear simultaneous equations, and the relaxation technique gives an orderly procedure for reducing to zero the 'residuals' at each node – in Hardy Cross's beams, for reducing to zero the out-of-balance moments at the joints.

6.5 Reciprocal theorems

The methods described in this chapter for dealing with problems of bending (beams and frames) were developed against the background of the establishment of a more general elastic theory of structures. Castigliano's theorems of 1879, for example, are applicable to frames as well as to trussed frameworks; they were in fact developed first for trusses, and will be discussed in Chapter 7. Application to trusses is reasonably straightforward; computations can be arranged in an orderly way, and techniques were devised for dealing with arrays of 'vectors' (a list of bar forces, for example, or of joint displacements). The application to bending problems was not so easy, since heavy algebraic manipulation is involved in the calculation of integrals.

However, the reciprocal theorems that apply to elastic structures gave fruitful results when used for the analysis of beams and frames. The fullest statement of the reciprocal theorem is that of Betti (1872), and the proof may be given simply, if anachronistically, by using the equation of virtual work. (Virtual work is exposed more fully in Chapter 9.) Figure 6.8 shows a simple frame, for which two statements will be made.

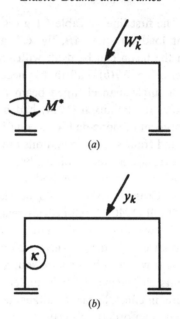

Fig. 6.8. (a) Bending moments M^* in equilibrium with external loads W_k^*; (b) curvatures κ compatible with displacements y_k.

In fig. 6.8(a) loads W_k^* are applied, and bending moments M^* in the frame are in equilibrium with those loads. In fig. 6.8(b) the frame has been distorted in bending, the curvatures κ giving rise to displacements y_k. Then the equation of virtual work states that

$$\Sigma W_k^* y_k = \int M^* \kappa \, dx. \tag{6.11}$$

The essential feature of this equation is that there is no necessary connexion between the equilibrium statement (W_k^*, M^*) and the geometrical statement (y_k, κ).

The bending moments M^* will, for an elastic frame, give rise to curvatures κ^*, where $M^* = EI\kappa^*$, so that equation (6.11) becomes

$$\Sigma W_k^* y_k = \int EI\kappa^* \kappa \, dx. \tag{6.12}$$

If now the roles of the two sketches of fig. 6.8 are interchanged, then, using the same arguments,

$$\Sigma W_k y_k^* = \int EI\kappa\kappa^* \, dx. \tag{6.13}$$

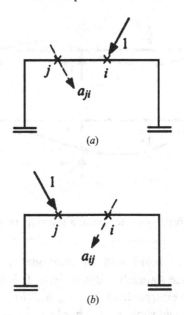

Fig. 6.9. Maxwell's reciprocal theorem: $a_{ij} = a_{ji}$.

Comparison of equations (6.12) and (6.13) leads to Betti's reciprocal theorem:

$$\Sigma W_k^* y_k = \Sigma W_k y_k^*. \tag{6.14}$$

In words, if two states (one starred and one unstarred) of an elastic body are considered, then the work done by the loads W_k^* of the first state on the displacements y_k of the second state is equal to the corresponding work done by W_k on y_k^*.

Maxwell's reciprocal theorem (to be stated in Maxwell's original form in Chapter 7) follows at once from equation (6.14). In fig. 6.9(a) a unit load applied in a specified direction at section i of a frame produces an elastic deflexion a_{ji} in a specified direction at section j of the frame. Similarly, fig. 6.9(b), an elastic deflexion a_{ij} results at section i from the application of a unit load at section j. Insertion of these statements into equation (6.14) gives Maxwell's result at once:

$$a_{ij} = a_{ji}. \tag{6.15}$$

Müller-Breslau's principle (1883) results from a similar application of equation (6.14). Figure 6.10(a) illustrates Müller-Breslau's own example

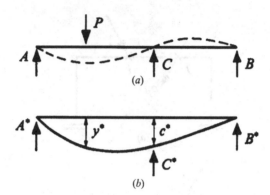

Fig. 6.10. Müller-Breslau's discussion of a statically indeterminate beam.

of a simply supported beam with an additional internal support, the system being thus once statically indeterminate. It is required to find the reaction C due to the external load P. In fig. 6.10(b) a small displacement c^* of the supposedly rigid support at C has ben imposed on the otherwise unloaded beam, inducing reactions A^*, B^* and C^* at the three supports (as sketched in fig. 6.10(b), the reaction C^* will have a negative value). If equation (6.14) is applied to the two states sketched in fig. 6.10, then

$$(A)(0) + (P)(y^*) + (C)(-c^*) + (B)(0)$$
$$= (A^*)(0) + (C^*)(0) + (B^*)(0), \qquad (6.16)$$

that is,

$$C = \frac{y^*}{c^*}P. \qquad (6.17)$$

Thus for an arbitrary unit displacement $c^* = 1$ of the internal support, and for a unit load P, the value of the reaction C at the internal support is equal to y^*.

There is therefore an immediate application to the construction of (elastic) influence lines. As a unit load crosses the beam of fig. 6.10(a), then the plot of the reaction at C – the influence line for the reaction at C – is given to some scale by the deflected shape of fig. 6.10(b).

6.6 Indirect model tests

Müller-Breslau's principle may be used to determine internal forces in an elastic frame. In fig. 6.11, for example, an imaginary arbitrary (unit) 'kink' has been introduced at the internal support of the same beam, in order

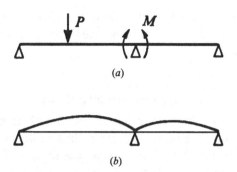

(a)

(b)

Fig. 6.11. Extension of Müller-Breslau's principle.

to determine the value of the bending moment M at that support which results from the application of the load P. Once again, the sketched deflected form of the beam in fig. 6.11(b) gives the influence line; that is, fig. 6.11(b) is a plot to some scale of the value of the bending moment M against the position of the point load P.

The right-hand side of equation (6.16) is zero since the supports in the original real beam of fig. 6.10 are rigid. Any system of deformation similar to that of fig. 6.10(b), for which imaginary displacements are introduced at supports that are in fact rigid, or similar to that of fig. 6.11(b), in which an imaginary internal dislocation is imposed, will lead to an equation of the form

$$\Sigma W_k y_k^* = 0, \tag{6.18}$$

cf. equations (6.14) and (6.16). Since equation (6.18) is homogeneous in the starred deflexion components, it would be possible to make the imaginary displacements, not on the real structure, but on a scale model having the same flexural characteristics as the real structure. All that is required is that the scale model should have flexural rigidities that are the same constant proportion from section to section as those of the original.

Beggs (1927) proposed that, instead of imaginary deformations, real deformations should be imposed on a carefully made and properly scaled celluloid model. Such a model can be cut from a plastic sheet of uniform thickness, the depths of the members being varied to ensure correct values of the flexural rigidities. The required coefficients (e.g. y^* and c^* in equation (6.17)) can then be obtained experimentally; internal discontinuities can be introduced to determine thrust, shear and bending moment at a cross-section. Observations of this kind can be very accurate, and acceptable estimates are obtainable even from beams and frames cut from cardboard.

6.7 General structural theory

The developments summarized in this chapter have been concerned with one type of structure, the beam (or frame) which resists external loads by the bending of its members. These developments took place, of course, within the evolving general theory of (elastic) structures, but their main concern was with the construction of ingenious methods for the approximate solution of large sets of linear simultaneous equations (Hardy Cross's method of moment distribution, for example), or for obtaining experimental solutions (Beggs's indirect model tests). This kind of activity was brought to an end by the invention of the electronic computer. The heavy labour of obtaining solutions to the equations could be undertaken by machine, and approximate solutions were no longer needed.

Matrix algebra had of course been applied to the solution of sets of linear equations, but in the 1950s specific attention was given by structural engineers to the formulation of structural theory (essentially elastic structural theory) in matrix terms. A leading figure in the development of this approach was Argyris; he gives a comprehensive summary in his review of 1954–55, where the work is strongly directed to the solution of aircraft structures. Livesley was concerned with early developments, and his book (1964) presents the theory with respect to beams and framed structures, and also to more general continuum problems in which the 'structure' is modelled by finite elements.

The construction of a general elastic theory, applicable to any structural type rather than the beams and frames considered in this chapter, emerged from a consideration of the trussed framework. As usual, Navier (1826) provides a convenient starting point for a description of contributions to the theory.

7

The Trussed Framework

7.1 Navier

In section 4 of Navier's 1826 *Leçons* he tackles the problem of the redundant truss. The example used is that of a weight Π supported from the ground by a number of bars, fig. 7.1, and the problem is to determine the forces in the bars. Navier states that if the number of bars is more than two in the same plane, or more than three not in the same plane, then the equations of equilibrium do not determine the values of the bar forces. Navier shows how the problem may be solved, using the three-bar plane-truss example of fig. 7.1.

There is first a short digression in which Navier attempts to estimate limits within which the bar forces must lie. If, for example, all bars are removed from a plane truss of the type sketched in fig. 7.1, except for the two necessary to carry the load, then the forces in those two bars may be found from the equations of statics. By considering different arrangements of bars to produce such statically determinate trusses, a greatest load may be found for a particular bar; the stability of that bar against buckling may then be checked, using the 'Euler' theory of a previous section of the *Leçons*. (These observations are, of course, incorrect. Even in the absence of the load Π, a turnbuckle tightened in bar $A'C$ will produce compression in the two outer bars, and buckling of one or the other will eventually occur. The ability of a redundant truss to sustain self-stress was certainly known to Maxwell (1864).)

Navier then lays out clearly the three groups of equations required for the elastic solution of the problem. The equilibrium equations, for the three-bar truss of fig. 7.1, are

$$\left. \begin{array}{r} p\cos\alpha + p'\cos\alpha' + p''\cos\alpha'' = \Pi \\ \text{and} \quad p\sin\alpha + p'\sin\alpha' + p''\sin\alpha'' = 0, \end{array} \right\} \tag{7.1}$$

111

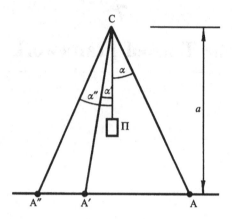

Fig. 7.1. Three-bar truss (after Navier (1826)).

where p, p' and p'' are the compressive forces in the three bars. No more equilibrium equations can be written; the truss has a single redundant bar.

Progress is made by Navier by writing a second group of equations, the compatibility equations. Navier supposes that, under the action of the load, the point C moves very small distances h and f, horizontally to the right and vertically downwards. Then the changes in lengths (shortening) of the bars compatible with these displacements are

$$\left.\begin{array}{c} f\cos\alpha - h\sin\alpha, \\ f\cos\alpha' - h\sin\alpha', \\ \text{and} \quad f\cos\alpha'' - h\sin\alpha''. \end{array}\right\} \qquad (7.2)$$

Since the lengths of the bars are $(a/\cos\alpha)$ etc, then the strains in the bars are

$$(f\cos^2\alpha - h\sin\alpha\cos\alpha)/a$$

and so on.

Finally, these elastic strains may be related to the bar forces by a third group of equations, those involving the 'elastic forces' F of the bars (where in usual modern notation $F = EA$, the area of the bar being A and Young's modulus E). Thus

$$\left.\begin{array}{c} p = F\left(f\cos^2\alpha - h\sin\alpha\cos\alpha\right)/a, \\ p' = F'\left(f\cos^2\alpha' - h\sin\alpha'\cos\alpha'\right)/a, \\ \text{and} \quad p'' = F''\left(f\cos^2\alpha'' - h\sin\alpha''\cos\alpha''\right)/a. \end{array}\right\} \qquad (7.3)$$

Equations (7.1) and (7.3), five in number, now give sufficient information to find the values of the three bar forces and the two displacements.

It may be noted that Navier keeps the equations completely general (e.g. three 'elastic forces' F, F' and F'') and also completely symmetrical – that is, he writes all three of expressions (7.2) for example, when in fact any one of the three conveys the necessary information. Further, Navier attaches a sign to the values of α, α' and α''; angles are positive clockwise measured from the downward vertical at C. Thus, if Navier's fig. 7.1 is interpreted literally, and numbers are inserted in equations (7.1) to (7.3), the sign of terms containing $\sin \alpha$ (but not $\sin \alpha'$ or $\sin \alpha''$) must be reversed.

7.2 Maxwell 1864

Figure 7.1 represents just about the simplest structural problem involving a trussed framework: there are only three bars, and the truss has a single redundancy. Even so, the analytical work is already heavy, and it increases enormously for a practical structure with many more bars and with perhaps several redundancies. For Navier's problem, in order to determine the three bar forces it was necessary to introduce two further unknown quantities (the components of the joint displacement), together with the three unknown bar extensions. Thus, in total, eight expressions, (7.1), (7.2) and (7.3), were formulated, in order to solve the prime structural problem, that of finding the three bar forces. The analytical methods for the solution of truss problems that were developed during the nineteenth century sought ways to reduce the number of unknown variables (in the same way that economy was sought in the solution of corresponding beam and frame problems, Chapter 6).

The history of the truss problem has been traced by Charlton (1982). Navier does not give 'priority' to forces or displacements in setting up his equations. However, his method, of introducing unknown displacements of joints in terms of which the (elastic) equations for the bars could be formulated, leads directly to what is now known as the equilibrium approach, or the displacement method. The formulation of these equations involves a more or less complex analysis of the geometry of the truss (the results of which are recorded, for example, in expressions (7.2) above). The use of the equation of virtual work transforms this problem from one of analysis of deformation to a simpler one of statics.

Maxwell (1864) used the equation of virtual work, although his analysis was confined to elastic trusses; his statement and proof of the method

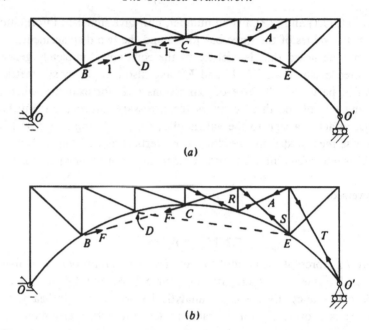

Fig. 7.2. Trusses to illustrate Maxwell's arguments (from Charlton (1982)).

are therefore restricted, although his technique is powerful. Instead of working with unknown displacements (Navier), Maxwell selects bar forces as unknown quantities whose values are required for the solution of a hyperstatic structure; equations of compatibility are written in terms of these 'redundant' quantities, and the method is now known as the compatibility approach, or the force method.

As a background to his work, Maxwell had knowledge of Clapeyron's theorem, as expounded by Lamé (1852). Clapeyron published his work formally in 1858, but his 'theorem' had been formulated by him some 30 years earlier, namely that the work done by the external forces acting on an elastic body is equal to the strain energy stored in the body. It was the brilliant way that Maxwell used this theorem that created the force method of analysis.

Maxwell published his 1864 paper without illustrations; fig. 7.2 copies Charlton's diagram to illuminate Maxwell's thought. He starts with a statically determinate truss, fig. 7.2(a), as a first step in the transformation of a problem of geometry to one of statics, and he makes the following highly condensed statement:

If p be the tension of piece A due to a tension-unity between the points B and C, then an extension-unity taking place in A will bring B and C nearer by a distance p.

This sentence records two distinct operations on the truss, and may be written, in more modern terms, as a statement of statics and a separate statement of geometry.

(i) Statics. (If) Equal and opposite unit loads applied at B and C in the line BC induce a tension p in bar A.

(ii) Geometry. (Then) As a separate matter, a unit extension of bar A (only) will cause points B and C to approach by a distance p.

A modern proof by virtual work might be presented as

(i) The tension X in A is in equilibrium with applied loads Y in the line BC.

(ii) An extension x of A causes an extension y of the line BC.

Then $Xx + Yy = 0$; but $X = pY$, so that $y = -px$.

This is almost exactly the proof given by Maxwell, except that he writes the work equation as $\frac{1}{2}Xx + \frac{1}{2}Yy = 0$, summing the *elastic* work. However, it may be noted that in Maxwell's condensed statement, and in both sets of statements (i) and (ii) above, there is no mention of material properties. Maxwell's statement is in fact completely general, and holds for an inelastic truss.

Maxwell then proceeds to a second subsidiary problem on the statically determinate truss, namely an examination of the relative movement of two joints D and E in fig. 7.2(*a*) caused by the application of a force F in the line BC. This force causes a tension Fp in bar A, and if the (elastic) 'flexibility' of the bar is e, then the bar extension is Fep. This is the geometrical statement to be used by Maxwell in (ii) below. The equilibrium statement arises from consideration of the effect of unit loads applied in the line DE; the resulting tension in bar A is q. Thus

(i) Statics. Equal and opposite unit loads applied at D and E in the line DE induce a tension q in bar A.

(ii) Geometry. A bar extension Fep in bar A leads to an extension of the line DE, say δ.

Then, by virtual work (as before), $\delta = -Fepq$.

If all the bars of the truss are considered, then the extension of the line DE is $-F\Sigma(epq)$; for the special case of line BC the extension is

$-F\Sigma(ep^2)$. Maxwell notes that p and q always enter the equations in the same way, so that a general (elastic) theorem may be established: the extension in BC due to a unity of tension along DE is equal to the extension in DE due to a unity of tension along BC. This is Maxwell's reciprocal theorem, which was stated in the shorthand form $a_{ij} = a_{ji}$, equation (6.15), fig. 6.9; the theorem is true for redundant as well as for statically determinate elastic structures.

These two 'lemmas' of Maxwell are prerequisites for his main problem, which is to find the bar forces in a statically indeterminate truss; Charlton's representation in fig. 7.2(*b*) shows redundant bars $R, S, T \ldots$. Maxwell starts by removing these redundant bars, reducing the truss to its statically determinate form, having tensions p and q in the typical bar A due to unit loads in the lines BC and DE, as before. Still using these lemmas, unit loads applied to the statically determinate truss in the lines of the (removed) bars R, S, T... will produce tensions r, s, t in the typical bar A. If the actual tensions in the redundant bars are $R, S, T \ldots$, and their flexibilities are $\rho, \sigma, \tau \ldots$, then Maxwell states that

> the tension in A is $Fp + Rr + Ss + Tt + \ldots$
> the extension of A is $e(Fp + Rr + Ss + Tt + \ldots)$
> the extension of R is $-F\Sigma epr - R\Sigma er^2 - S\Sigma Ers - T\Sigma Ert - \ldots = R\rho$

Similarly, the extensions of S and T are given by

$$S\sigma = -F\Sigma eps - R\Sigma ers - S\Sigma es^2 - T\Sigma est - \ldots$$
$$T\tau = -F\Sigma ept - R\Sigma ert - S\Sigma est - T\Sigma et^2 - \ldots$$

and there are enough equations (three for three redundancies R, S and T) to determine the unknown quantities. Finally, the extension x of the line DE may be calculated from

$$x = -F\Sigma epq - R\Sigma eqr - S\Sigma eqs - T\Sigma eqt - \ldots \qquad (7.4)$$

It may be noted that if point E (say) is actually a fixed point of the truss, and if this point is remote ('the centre of the earth'), then the value of x calculated from the equation (7.4) will give the (vertical) deflexion of D due to a load F applied at that point.

7.3 Virtual work

It will have been seen that Maxwell's method, as outlined above, involves only statics for the solution of problems; the coefficients p, q and so on are determined from the consideration of equilibrium, and the geometry of deformation does not have to be tackled directly. The method is really one of virtual work, although this is disguised by the fact that Maxwell considers only linear elastic systems. Fleeming Jenkin (1869) extended Maxwell's method with a more exposed use of 'virtual velocities', although once again he analysed only elastic trussed frameworks.

The particular application considered by Jenkin may be illustrated by reference to fig. 7.2(*a*), in which the abutment $0'$ is now considered to be rigid, and hence capable of sustaining a horizontal abutment thrust H. Indeed the value of H may be considered as the single redundancy of the problem, whose value is required as the result of the application of a specified loading system. Jenkin imagines the truss to be freed from horizontal restraint at $0'$; a unit horizontal load applied between the abutments will give rise to a bar force q in a particular bar. Again, an extension x of the particular bar, considered in isolation, may be related to the virtual horizontal displacement y of $0'$ with respect to 0, since, by virtual work,

$$1 \cdot y = qx. \tag{7.5}$$

Now the force in the typical bar under the specified loading (say V, where V is representative of all the applied loads) may be written as $(pV + qH)$, so that the bar extension is

$$x = e(pV + qH), \tag{7.6}$$

that is,

$$y = e(pqV + q^2H). \tag{7.7}$$

Equation (7.7) is written similarly for all bars of the truss; the summation Σy must be zero, since in reality the abutments do not move relative to each other. Thus

$$H = -V\frac{\Sigma epq}{\Sigma eq^2}. \tag{7.8}$$

The work of Maxwell and Jenkin did not at first receive wide attention; it was unknown to Mohr (1874), who developed an analysis along very similar lines. Mohr's work was, by contrast, widely accepted, and after Mohr's acknowledgement of Maxwell's priority, engineers described the analysis as the Maxwell–Mohr method.

7.4 Energy methods

Maxwell cited 'Clapeyron's energy theorem', but antecedents can be traced to at least a century earlier. It was seen (Chapter 4) that the Bernoullis, and Euler, had used the idea of strain energy in bending to solve the problem of the shape of the elastica. However, it was Castigliano (1879) who gave a full formulation of the use of energy principles as a general method for finding solutions for elastic structures. In fact, unknown to Castigliano, some of this work had been anticipated by Cotterill (1865); Charlton traces thoroughly the various contributions in the second half of the nineteenth century.

Castigliano develops the theorems that bear his name with reference to trussed frameworks (he extends the work later to cover beams and arches). He first formulates expressions for the internal strain energy W_i in terms of the applied loads F_p acting at the various nodes of the truss; at these nodes the displacements are r_p (measured in the direction of F_p). If a typical bar has tension P, length ℓ and 'elasticity' AE, then

$$W_i = \frac{1}{2}\Sigma F_p r_p = \frac{1}{2}\Sigma \frac{P^2 \ell}{AE}. \tag{7.9}$$

Then Castigliano's first theorem, part 1, is

$$\frac{\partial W_i}{\partial r_p} = F_p, \tag{7.10}$$

and part 2 of the same theorem is

$$\frac{\partial W_i}{\partial F_p} = r_p. \tag{7.11}$$

Castigliano's proof of equation (7.10) is straightforward. He imagines the force F_p to be increased by an amount dF_p, resulting in an increase dr_p in the displacement r_p of the node. Then (neglecting products of infinitesimal quantities), the increase in total internal work will be

$$dW_i = \Sigma F_p dr_p; \tag{7.12}$$

but since

$$dW_i = \Sigma \frac{\partial W_i}{\partial r_p} dr_p, \tag{7.13}$$

and since equations (7.12) and (7.13) must hold for arbitrarily selected values of dr_p, then equation (7.10) follows.

Similarly, if W_i is considered as a function of the applied forces F_p,

$$dW_i = \Sigma \frac{\partial W_i}{\partial F_p} dF_p = \Sigma F_p dr_p. \tag{7.14}$$

But from equation (7.9),

$$dW_i = \frac{1}{2}\Sigma r_p dF_p + \frac{1}{2}\Sigma F_p dr_p, \tag{7.15}$$

and, using equation (7.12),

$$\Sigma F_p dr_p = \Sigma r_p dF_p. \tag{7.16}$$

Thus equations (7.14) and (7.16) combine to give

$$\Sigma \frac{\partial W}{\partial F_p} dF_p = \Sigma r_p dF_p, \tag{7.17}$$

and, as before, equation (7.11) follows.

Castigliano's second theorem, the theorem of least work, follows easily from part 2 of the first theorem. If the truss has a number of redundant members (say $R, S, T \ldots$ following Maxwell's notation), then the theorem states that the values of the forces in those bars, $R, S, T \ldots$ say, will be such that the internal strain energy is a minimum, that is

$$\frac{\partial W_i}{\partial R} = \frac{\partial W_i}{\partial S} = \frac{\partial W_i}{\partial T} = \ldots = 0. \tag{7.18}$$

To prove the theorem, Castigliano removes a redundant bar connecting two nodes, say bar R, and replaces it by equal and opposite forces R acting at those nodes. The expression $\partial W_i/\partial R$ for this modified truss then gives, from equation (7.11), the relative displacement of the nodes, and this must be exactly equal to the extension of bar R, namely $R\ell/AE$. Now this expression is also given by $\partial W_i/\partial R$ from equation (7.9), and the first of equations (7.18) for the original unmodified truss then follows immediately. There are, of course, exactly as many equations (7.18) as there are unknown (redundant) quantities whose values are to be found.

An essential feature of this proof is that the truss, before loading, consists of bars each of exactly the correct length. If a bar or bars are of incorrect initial length, then the (redundant) truss will be in a state of self stress, and Castigliano shows how to calculate this state from the known values of misfit (including misfits due to differential temperature changes).

The term strain energy was used by Castigliano to denote the internal energy stored in a linear-elastic system. Engesser (1889) showed that equation (7.11) still held for non-linear systems provided that W_i was interpreted as the complementary energy (the area shaded in the

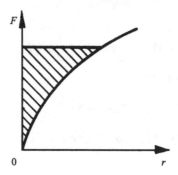

Fig. 7.3. Non-linear force/extension curve for a truss member; complementary energy is represented by the shaded area.

force/extension diagram sketched in fig. 7.3), i.e.

$$W_i = \int r dF. \tag{7.19}$$

However, by far the greatest use of Castigliano's theorems was for the solution of linear-elastic structures, and the more general application of virtual work was ignored. Maxwell used virtual work in an elastic context; Fleeming Jenkin had a broader view, although he still dealt with elastic structures; Mohr and Engesser were well aware of the idea. It seems unlikely that Castigliano appreciated the use of virtual work as a fundamental tool – certainly his theorems can be proved more simply and more elegantly by its use.

8

Scholium 1

Galileo's concern was with the breaking strength of a cantilever beam. The behaviour of such a structure is determined by the equations of statics and by the strength of the material; there is only one internal force system in equilibrium with the applied loads and, for the bending problem, collapse will occur when the value of the largest internal bending moment reaches the moment of resistance of the cross-section. Thus the problem of finding the actual state of a statically determinate structure and the problem of calculating its strength are, effectively, one and the same.

This is, of course, not so far the hyperstatic structure. Historically, three types of hyperstatic structure were examined (and the theories have been described in previous chapters) – the (masonry) arch, the continuous beam and the trussed framework. It is of interest that the early (eighteenth-century) work on arches did not concentrate on the 'actual' state – rather, limiting states were examined in order to determine the value of one of the main structural parameters, the abutment thrust. This approach continued through the nineteenth century until Castigliano applied his elastic energy theorems to both iron and masonry arches in order to calculate the same structural parameter. Thereafter, arch analysis was seen to fall within the mainstream techniques for the elastic design of hyperstatic structures.

Similarly, specialized elastic techniques were developed for redundant beam systems. Statics alone did not give enough information; the second-order differential equation of bending introduced the elastic properties of the sections; and the boundary conditions (clamped ends, rigid supports) provided the geometrical information leading finally to sufficient equations to solve the problem. These general methods led to compact analytical techniques (for example, the theorem of three moments, which is an example of the compatibility approach, in which the redundant quantities are taken as unknown bending moments), and later on to techniques of numerical computation (for example, Hardy Cross's method of moment distribution, which is an example of the equilibrium approach).

As has been seen, the force method (the compatibility approach) was exploited for the elastic solution of the trussed framework. Once the forces in the redundant bars had been determined – directly by virtual work, or by using Castigliano's energy theorems – then the forces in all the bars could be found, and the actual state of the structure determined. Progress towards a general theory of structural analysis was made to a great extent with respect to trusses, and perhaps the main reason for concentration on this type of structure was its increasing and widespread use in railway engineering. A technical reason was mentioned in Chapter 6, namely that the vectors of bar forces, bar extensions and joint displacements could be assembled and manipulated in an orderly way. By contrast, the differential equations involved in the discussion of beams and frames could lead to heavy mathematical labour.

Whatever the type of structure, whether truss or beam, the *purpose* of analysis was clearly defined by Navier in 1826 – it was to determine the values of the internal stress resultants, whether bar forces or bending moments, and to ensure, in the first instance, that the sections provided gave adequate strength. (Other structural criteria were not neglected; deflexions could be calculated, and 'Euler' stability could be checked.) With the objectives stated in this way, no thought seems to have been given to the uniqueness of the solutions for hyperstatic structures. That is, a statically determinate structure has a straightforward solution; bar forces in a truss resulting from the applied loads can be found at once. Similarly, the structural techniques of Navier and Maxwell determine the values of bar forces for a statically indeterminate truss; the solution is not straightforward, but nevertheless values of the bar forces result.

Once this 'actual' state of a structure, whether statically determinate of hyperstatic, had been found, then Navier was again quite clear how the strength check should be made – the stresses at any section of the structure should not exceed a certain proportion of the limiting stress of the material (yield or fracture), that is, the material should remain elastic. There is no hint in Navier's work that some initial state of self-stress in a redundant truss might lead to different final values of the bar forces.

8.1 Secondary stresses

In the same way, although Castigliano, half a century later, was well aware that initial lack of fit of bars in a truss could lead to substantial values of bar forces, he assumes in his worked examples of real structures that no such lack of fit occurs, and seems untroubled by any thought that imperfections of manufacture or construction might invalidate his calculations. On the other hand, he shows how to calculate stresses due to a differential rise in temperature – stresses which are, of course, exactly equivalent to those arising from a practical imperfection.

Thus, in his worked example of an iron truss of 48 m span, which has three statical indeterminacies because of its attachment to rigid abutments, he assumes explicitly that the truss was assembled at 15 °C without prestress; he calculates the primary forces resulting from dead and superimposed loading of 126 kg/m^2, of which snow load forms approximately one half. Castigliano then examines the effect of variation in temperature between -10 and +40 °C, the abutments remaining rigid. The calculations are done separately, that is, Castigliano determines a self-stressing system of bar forces in the absence of external load and, since the equations are linear, this system of 'secondary' forces may be imposed ('by the principle of superposition') on the primary forces calculated previously. The resulting bar forces turn out to be quite large, although Castigliano observes that there would be no snow load if the temperature were 40 °C.

The term 'secondary stresses' usually denotes a different set of self-equilibrating stresses which can arise in a structure, and in particular in trusses. From the time of Navier and Maxwell the forces in the bars of a truss were calculated on the assumption that the members were pin-ended, so that purely axial forces arise. The bars in a real truss are connected to each other by gusset plates and a number of bolts or rivets, making the joints more nearly rigid than pinned. The pin-ended calculations give primary forces from which primary stresses may be determined; on these stresses must be imposed the 'secondary' effects due to the rigidity of the joints. Charlton gives a brief account of the development of this sort of calculation, starting with Asimont (1880), who coined the terms. Several famous names are associated with the work, including Manderla (1880), Engesser (1879 and 1892), Winkler (1881), Müller-Breslau (1886) and Mohr (1892).

The first analysis of the pin-jointed truss gives the primary forces, and also the displacements of the joints. The real joints of the stiff truss are then assumed to occupy these same displaced positions, but this involves bending of the bars in order to maintain compatibility of rotation at each of the rigid connexions. Thus, in Manderla's work, the 'slope-deflexion equation'

$$\phi_1 = \frac{\ell}{6EI}(2M_1 - M_2) \qquad (8.1)$$

emerges, involving the rotation ϕ_1 (with respect to the chord) of one end of a member when subjected to couples M_1 and M_2 at the two ends (cf. equation (6.8), written with a different sign convention); a similar equation is written for the rotation ϕ_2. Mohr, on the other hand, inverts these equations, to give

$$M_1 = \frac{2EI}{\ell}(2\phi_1 + \phi_2) ; \qquad (8.2)$$

he can then write conveniently the equilibrium equation for the joint as $\Sigma M_1 = 0$, where the summation is taken over all members meeting at the joint. Exactly enough equations can be written to determine the values of these secondary bending moments, and hence the secondary stresses can be calculated. It may be repeated that these secondary stresses, like Castigliano's temperature stresses, are self-balancing; they are in equilibrium with zero external load applied to the truss.

It will be appreciated that the number of equations involved in this analysis is very large; further, the primary calculation of the deflexions of the joints is also complex, and ingenious contributions were made to lighten the burden of this part of the analysis. The methods evolved (of course for hand calculation before the advent of the electronic computer) persisted for 50 years, and they formed a substantial part of the design process for large trussed bridges (for example, the Sydney Harbour Bridge, 1928–32; see Pain and Gilbert Roberts (1933–34)).

Although all this work was complex, the objective remained that of Navier in 1826. The total stress at the end of a bar of the truss was calculated from the primary axial force, and to this were added temperature stresses and secondary bending stresses; the total was not allowed to exceed a certain fraction of the yield stress. (Depending on the terms appearing in the final total, the fraction could be increased in sympathy with Castigliano's 'no snow at 40 °C' observation.) Once again, those making the calculations were not troubled by any thought that the stresses might be in any sense 'unreal' – that practical imperfections, of manufacture or assembly, might lead to an observable state different from that calculated. The fact was that, in general, no observations were made to confirm the correctness of the structural theory. That theory was self-evidently correct: the equilibrium equations were written properly; material parameters (elasticity and yield strength) were known; and the geometry of the structures could be described in a straightforward way. It seems simply not to have been appreciated that if a hyperstatic structure were built with only minute variations from the prescribed geometry, either internally or in its connexions to the foundations, then very large variations in stress would result.

8.2 Nineteenth-century experiments

An exception to this lack of experiment has been noted in Chapter 4; Euler's buckling formula was confirmed for slender columns. And other structural experiments were in fact made in the nineteenth century, but these provided solutions to problems for which little or no theory was available. Fairbairn, for example, had made many material tests in the 1830s, particularly on cast and wrought iron, and had developed a convenient testing machine (Fairbairn's lever). His experimental expertise was sought in 1845 by Robert Stephenson to assist with the solution of a structural problem, that of the design of the tubular spans proposed for the Britannia and Conway bridges (see Fairbairn (1849)). The tubes were large enough for trains to run inside them, and preliminary experiments had shown that failure occurred by compressive buckling of the thin plates forming the section. The extensive experimental programme led to efficient design rules for these tubular girders, including the provision of stiffeners to the plates. Construction of the Britannia bridge had started before these experiments were complete, and the masonry towers between which the tubes spanned had been built sufficiently high that the tubes could be given additional support by suspension chains. However, the final design was strong enough for these chains to be omitted, the tubes acting as beams carrying their own weight and the superimposed loads.

Jouravski (1860), whose work on shear stresses in beams was noted in Chapter 3, also made buckling experiments, and as a result had comments to make on the design of the Britannia bridge. For example, the stiffeners to the sides of the Britannia tubes were vertical, whereas the maximum compressive stresses causing buckling acted at 45° – the stiffeners would be better employed at that angle. Further, Jouravski seems to be the first engineer to test specimens not made of the same material as the real structure; elastic buckling is more easily observed in materials having small elastic moduli, and Jouravski used thick paper and cardboard to achieve this end. All such nineteenth-century experiments, however, were made in areas of structural engineering for which mathematical theory was lacking – where a problem had been 'solved' (the solution for the continuous beam, the calculation of bar forces in a redundant truss) there was apparently no thought of making confirmatory tests.

Evidence from the previous two centuries is mixed. There is no record of bending tests made by Galileo, but Mariotte (1686) did make tests on simply supported beams and he produced accompanying theory; his concern was to establish correct values for the ultimate moment of resistance. As such, these tests are not really structural; their concern is with a problem in strength of materials. Similarly, the nineteenth-century experiments (Eaton Hodgkinson (1824, 1831) and others referred to in Chapter 2) were directed to the solution of 'Galileo's problem' and not to the advance of structural theory. The fixed-ended beam, on the other hand, is a problem in the theory of structures; but when Mariotte tested such beams, and established that they could carry twice as much load as the corresponding simply supported beams, he did not support this result with any theory. Girard (1798) attempted some analysis, but he reported no experiments. It would seem that over a hundred years more were to pass before any tests were made to support (or disprove) established theory; it was in 1914 that Kazinczy published (in Hungarian) his 'Tests with clamped beams'.

9

Plastic Theory

9.1 Kazinczy

Kazinczy (1914) tested two steel beams, each about 6 m long, which were embedded at their ends in substantial abutments; the loading, which consisted of increasing numbers of courses of bricks, was uniformly distributed. The steel beams were in fact encased in concrete, but Kazinczy easily dissects out the conclusions that apply to the steel alone. If the ends of the beam in fig. 9.1(a) were perfectly fixed, then conventional elastic theory gives the bending-moment diagram sketched in fig. 9.1(b); the beam must be designed for a maximum bending moment of value $w\ell^2/12$. The explicit question asked by Kazinczy, to which the experiments were designed to provide the answer, was whether the end embedment may be taken to be complete and, if not, what degree of fixity may be assumed.

The concrete provided an effective tell-tale to monitor the progress of the experiments. As the loading was increased, cracks in the casing first appeared at the ends of the beams, indicating yield at those points. However, the beams could carry further load, and it was not until a substantially greater weight had been added that deflexions became very large. Upon unloading, each beam was found to have permanent kinking deformation, at the two ends and at the centre. Kazinczy called these kinks 'hinges', and he states that a fixed-ended beam cannot collapse (undergo increasing deflexions) until three hinges have formed. Two (end) hinges merely transform the fixed-ended into an effectively pin-ended beam; the third central hinge is necessary for collapse. Moreover, says Kazinczy, the degree of end clamping is irrelevant, provided the embedment is strong enough to allow the hinges to develop. Thus a

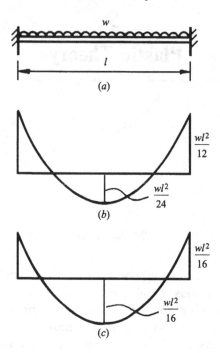

Fig. 9.1. Collapse of a fixed-ended beam under uniformly distributed load.

crucial, if surprising, answer has been given to the question for which the experiments were designed.

As a result of these tests, Kazinczy concludes that such beams could be designed for a largest bending moment of value $w\ell^2/16$, fig. 9.1(c). Indeed, he states that the moment is $\frac{1}{2}(w\ell^2/8)$; the carrying capacity of a fixed-ended beam is twice that of the equivalent simply supported beam. It is clear that Kazinczy is thinking of $w\ell^2/8$ as the value of the 'free' bending moment for an equivalent simply supported beam, and that the carrying capacity for the fixed-ended beam can be found by superimposing a suitable base line on the free bending-moment diagram.

9.2 The 1936 Berlin Congress

Progress on plasticity theory, as it came to be called, was resumed in the 1920s, after World War I, mainly in central Europe (Germany, Poland, Austria) and also to some extent in Switzerland and France. The International Association for Bridge and Structural Engineering (IABSE) held a first Congress in Paris in 1932; by the time of the second Congress, in Berlin in 1936, eight papers in the general field of plasticity formed one section of the proceedings. Some of these papers are non 'structural' – that is, they are concerned with the strength-of-materials aspect of the formation of plastic hinges; Galileo's problem is re-examined in the light of non-linear behaviour (seemingly in ignorance of Saint-Venant's exploration of the subject). More generally, the collapsing structure is examined in a 'historical' way; that is, the *elastic* solution is studied and modified as the loading is increased on the structure. It is, of course, not surprising that engineers in the 1930s should start from a conventional elastic approach in attempts to describe the actual behaviour of structures.

As an example of this approach, Kazinczy himself provided a summary paper in the final Proceedings of the 1936 Congress. He comments on the contributions from Maier-Leibnitz and by Melan, discussed below, and he states that it is a major achievement of plastic theory to conclude that imperfections, such as settlements of supports or residual stresses due to manufacture, may be ignored in the calculation of collapse loads. However, in describing the way in which this calculation is done, he states that bending moments must first be determined according to elastic theory, and then the base line may be shifted in order to equalize the largest moments. This notional procedure is of course based on the behaviour of structures when actually loaded to collapse.

Kazinczy is more penetrating in his discussion of allied matters – for example, the need to increase the size of members if there is danger of instability. Or again, it was clear in 1936 that if a framed structure had a number n of statical indeterminacies, then a number n of hinges would make the structure determinate, and $(n + 1)$ hinges are required for collapse. Kazinczy superimposes on this the need for the hinges to be of the right signs, so that a kinematic chain is possible, and these ideas embrace the possibility of partial collapse. Indeed, Kazinczy foreshadows the weak-beam/strong-column philosophy of design of structural frames – the beams may be designed plastically to have minimum sections, while the columns are deliberately strengthened to guard against buckling.

9.3 Maier-Leibnitz

Maier-Leibnitz, of Stuttgart, knew of Kazinczy's early work, and his own paper in the 1936 IABSE proceedings reports a large number of tests on continuous beams; the paper reached significant conclusions and, as will be seen, was influential in the development in the United Kingdom of plastic methods for the design of steel structures. He defines what is meant by 'carrying capacity', distinguishing between the attainment of yield at a single cross-section and the development of large (ductile) deflexions at a higher load (as had already been noted by Kazinczy). In one particular series of tests, Maier-Leibnitz loaded to collapse three beams, each 4.8 m long, supported at their ends and also centrally. In the first test the three supports were level. In the second, the central support was lowered (before the application of the central load) to a point where the bending stress at the support had the yield value; the beam was then loaded to collapse. In the third test the central support was raised by the same amount, again generating the yield stress, and the beam was again loaded to collapse. In this third test, conventional elastic theory would prohibit the addition of any external load. The actual collapse loads in the three tests (that is, the loads at which deflexions became large) were 13.1, 13.0 and 13.45 tonnes respectively. These three tests (and Maier-Leibnitz quotes others, made by himself and by other workers) confirm that collapse occurs when hinges form in sufficient numbers to form a mechanism, and provide a demonstration that the collapse load is essentially unaffected by initial imperfections, such as sinking of supports.

Maier-Leibnitz sketches bending-moment diagrams in the way shown in fig. 9.2 (illustrating the three tests described above), that is, they are combinations of 'free' and 'reactant' diagrams. (It should be remarked that this construction for bending-moment diagrams was proposed by Bertot (1855), and described so carefully as to make it likely that the construction originated with him.) The single redundancy has been thought of as the bending moment at the central support; with this set to zero, the free moments for the two simply supported spans may be drawn. Superimposed on these moments is the reactant line arising from the presence of the actual moment over the central support. Although the diagrams are drawn in this way, however, the reactant line is placed initially as the result of an elastic analysis, and the value of the central moment is then traced as the external load is increased and plasticity develops.

Maier-Leibnitz's paper gives a comprehensive bibliography, and it is of interest that the name of W. Prager is mentioned, but not referenced; Prager's work is discussed below. Also of interest is that one collapse test on a portal frame is noted, by K. Girkmann (1932); this seems to be the first such test reported in the literature. The pin-based frame, carrying a single central vertical load, collapsed by the formation of three hinges

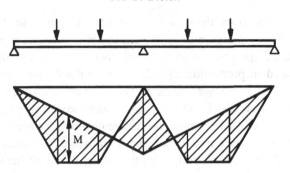

Fig. 9.2. Bending-moment diagram for two-span beam, after Maier-Leibnitz (1936).

in the beam; again the analysis is carried out 'historically', by recording the movement of the 'thrust line' from its elastic position as the load is increased.

9.4 F. Bleich

Maier-Leibnitz also gives three references to the work of F. Bleich, and Bleich contributes another paper to the 1936 IABSE proceedings. This summary paper is concerned to a great extent with the plastic behaviour of beams and frames when they are subjected to a specified range of alternative loading conditions; the presentation leads on to the phenomenon of shakedown, as will be seen. However, Bleich starts the paper with statements of a number of concepts crucial to plastic methods.

First, he abandons the idea of an elastic factor of safety on stress. Instead, he introduces the idea of *load factor*, defined as the ratio of the collapse load to the specified design working load. It is assumed tacitly that all loads acting on a structure have applied to them the same load factor. In later descriptions, this is known as *proportional loading*; all loads are imagined to be increased by the same multiplying factor λ, and collapse occurs when λ reaches the ultimate value λ_c.

Second, Bleich defines the type of material needed for a plastic theory of design: an elastic stress–strain relationship must be followed by indefinitely large deformation at the plastic limit, that is, the material must be ductile. If this ideal plastic limit is reached at a yield stress σ_0, then the full plastic moment of a section in bending will be $M_p = z\sigma_0$, where z is the value of the (plastic) section modulus.

It is apparent from these two statements that, in 1936, the plastic method of design was seen to arise from the idea of a (beam or framed) structure acted upon by certain specified working loads, with those loads then increased in proportion by a factor λ until collapse occurs by the formation of plastic hinges, sufficient in number to turn the structure into a mechanism. Bleich, however, saw much deeper into the method, and made an alternative formulation which obviated the need to imagine a hypothetical increase in load leading to collapse. His calculations are carried out (as for elastic design) with *working* values of the loads, but with the value of the yield stress (and hence proportionally the value of the full plastic moment) reduced to σ_0/λ_c. As he points out, this will ensure the attainment of the required reserve of strength expressed by the load factor λ_c.

Third, he makes use of the fundamental property of hyperstatic structures, namely that they are capable of self-stress in the absence of external load. Moreover, he notes that such states of self-stress can arise from first loading a structure into the plastic state, and then unloading. He then states (what is now known as) the lower-bound theorem for the general case of a structure acted upon by variable and repeated loads. At each of the critical cross-sections of a beam or frame the greatest and least bending moments, $\mathcal{M}^{\max}$ and $\mathcal{M}^{\min}$, are calculated by the usual elastic theory for statically indeterminate structures. The structure will be capable of carrying the prescribed loads if self-stressing moments m can be found such that, at each cross-section,

$$\left. \begin{array}{r} \mathcal{M}^{\max} + m \leq M_\mathrm{p} \\ \text{and} \quad \mathcal{M}^{\min} + m \geq -M_\mathrm{p}, \end{array} \right\} \qquad (9.1)$$

where, of course, the required load factor λ can be applied either to the elastic moments $\mathcal{M}$ or inversely to the values of the full plastic moment M_p. Bleich does not prove this theorem, but refers to the proof by Melan (see below).

Further, and crucially, Bleich observes that elastic calculations are not necessary for the case of *fixed* loads (he actually envisages a wider class of loading, where the loads may vary but are always positive – the phenomenon of incremental collapse is not in consideration here). For these calculations it is not necessary to use the elastic moments $\mathcal{M}$, but merely to compute bending moments for an equivalent statically determinate structure. This leads to the kind of diagram sketched in fig. 9.2, in which free and reactant moments are superimposed in such a way that the net moment $|M|$ is less than M_p at every cross-section.

Fourth, Bleich notes that practical imperfections (e.g. settlements of supports) give rise to self-stressing moments, and that since the plastic solution requires the adjustment of the values of self-stressing moments (in order to satisfy equations (9.1)), then those practical imperfections can have no influence on the safety of a statically indeterminate structure. He identifies temperature variations as giving rise to similar self-stressing moments, but, since such variations may be repeated indefinitely, he is unwilling to ignore their influence. This is because he is not satisfied with the evidence on fatigue loading of plastic structures, although he concedes that fatigue fractures of building frames are 'scarcely known'. Until such evidence is available, Bleich advises that moments arising from temperature changes should be taken as contributing to $\mathcal{M}^{\max}$ and $\mathcal{M}^{\min}$ in a 'shakedown' analysis.

The second half of Bleich's paper is taken up with examples of practical calculation of three different cases of continuous beams, and of a pitched-roof fixed-base portal frame. He cautions specifically against the use of plastic methods for the design of lattice girders, since compression members tend to buckle and do not have the required structural ductility.

9.5 Melan

A third paper of great significance in the 'plasticity' section of the 1936 Berlin Congress was presented by E. Melan, of the Technische Hochschule, Vienna. The title is simply 'Theory of statically indeterminate systems' (*Theorie statisch unbestimmter Systeme*), and its two sections deal with structures made of ideally plastic material, and of material that exhibits linear strain-hardening. It is not easy to determine which parts of the very rich material presented originated with Melan, and which summarize common intellectual property in the 1930s. Melan refers to the early twentieth-century contributions of von Karman, von Mises and Hencky; equally, he is aware of the work of Fritsche, Grüning, Kazinczy, H. Bleich and F. Bleich, Hohenemser and Prager, and it is clear that these in turn were well aware of Melan's own work.

Thus the opening paragraphs of the paper may do no more than collect together common ideas of 1936, but the statements do not seem to have been made earlier with such precision and clarity. The first sentence confirms that the object of statics as applied to steel structures is to determine the internal stresses and deformations of systems composed of slender members. The 'theory of steel structures' defined in this way is therefore not a branch of continuum mechanics; 'slender members' imply that only a single stress resultant is being considered. For example, Hooke's Law applied to a beam or framed structure states that curvature is proportional to bending moment – applied to a truss, that elongation of a bar is proportional to axial load. Melan in fact presents his mathematics with respect to trussed frameworks (he repeats Bleich's warning about buckling), so that the use of integrals is avoided, and simple equations can be written, with the use of summation signs if needed (Melan employs a kind of summation convention used in tensor notation). However, he states that the results obtained with this less complex mathematics may be applied to bending problems.

The second sentence of Melan's paper notes that only three groups of equations are available for solving structural problems – equilibrium equations, geometrical equations (compatibility) and stress–strain relations. For statically determinate systems the equilibrium equations by themselves suffice to determine the internal stress resultants, but all three groups of equations must be used for indeterminate systems to solve the same prime structural problem. All this may appear to be, and indeed is, self-evident, but the statements of the purpose and techniques of structural analysis are particularly clear. For example, Melan notes that usual 'text-book' theory assumes elastic behaviour, but that this assumption has in fact no influence on the determination of stress resultants in statically determinate structures. By contrast, the calculations for hyperstatic structures give results that depend on the actual stress–strain relationship that is used in the analysis. Melan then gives a condensed but comprehensive survey of theories involving non-linear stress–strain assumptions, particularly that of the 'ideal-plastic' material sketched in fig. 9.3, where the axes may be stress and strain, or moment and curvature.

In particular, Melan notes that some of the properties of elastic systems are lost if non-linear behaviour is assumed. For example, the law of superposition is lost for all but the forces in statically determinate structures, so that, for example, influence lines cannot in general be constructed. Crucially, it is impossible to define the stress state of a statically indeterminate structure unless its previous loading history is known; if the elastic limit has been exceeded, in some unknown and undefined way, then an unknown state of residual stress will exist in a hyperstatic structure. It is precisely this state of residual stress which may enable a structure to 'shake down' under the application of specified random loading. Melan's proof of the shakedown theorem, expressed essentially by inequalities (9.1) above, is perhaps the major contribution he makes in his paper; he attributes the theorem to H. Bleich (1932), although Ble-

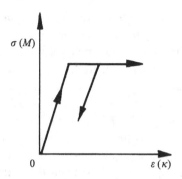

Fig. 9.3. Ideal-plastic stress–strain relationship.

ich gave only a limited proof. A simplified statement of the shakedown theorem is given in §9.10 below.

Finally, Melan quotes the principle of virtual work (and uses it in his proofs) which makes use only of the first two structural equations (equilibrium and geometry), and is not concerned with any statement of material properties. Written in terms of the trussed framework discussed by Melan,

Equilibrium: Internal bar forces T are in equilibrium with external loads W applied at the nodes of the truss.

Geometry: Bar elongations e are compatible with movements δ of the nodes of the truss.

These are two independent and unrelated statements for the truss under consideration. Then the principle of virtual work states that

$$\Sigma W \cdot \delta = \Sigma T e. \tag{9.2}$$

9.6 J.F. Baker

In the UK the application of plastic theory to the design of steel-framed buildings is associated with the name of J.F. Baker (Sir John Baker; Lord Baker). Steel frames had been in existence since the start of the twentieth century; they were designed on the basis of elastic theory, transmuted into codes of practice for the benefit of the hard-pressed engineer. There were many such codes in use throughout the world, and they contained conflicting rules; in 1929 the British steel industry set up the Steel Structures Research Committee to try to bring some order into practical steel design. The SSRC included eminent academics and officers from government research stations, and leading representatives from the consulting and contracting professions. Baker was the full-time technical officer, and it was his task to assemble technical information, to write or commission papers developing theory, and to oversee the collection of experimental evidence. The findings of the Committee are collected in the three volumes published in 1931, 1934 and 1936, and some of the work is summarized in volume 1 of Baker's *The Steel Skeleton* (1954).

These papers reveal the state of understanding, in the UK in 1936, of the structural design process, as indeed does Baker's own text-book (written with A.J. Sutton Pippard) published in that same year. Design and analysis are completely elastic processes, and the tools available include, for example, Maxwell's reciprocal theorem, the theorems of Castigliano and Hardy Cross's method of moment distribution. It was to be many years before the basic principles of virtual work crossed from Europe (via the US) to the UK, and standard texts in English, although learned, were in a sense somewhat rigid. An exception lies, perhaps, in the treatment of beam-columns, that is, of members subjected both to bending and to axial thrust. Baker's own papers in the SSRC volumes, and the corresponding chapter in his 1936 text, contain ingenious developments of solutions of the fundamental equations.

However, the outstanding contribution of the Steel Structures Research Committee to the question of structural design lay in experimental work. New steel buildings were being constructed in the 1930s, and the Committee arranged for tests to be made on (among other structures) a nine-storey hotel block, an office building and a block of residential flats. The results are reported in the three volumes published by the SSRC, and may be found also in volume 1 of *The Steel Skeleton*. For the first time stresses in real structures were measured (and the development of suitable strain gauges was an essential part of the work). In summary, the real stresses bore almost no relation to those calculated by designers using the available elastic methods. The SSRC were not slow to find the reason; small errors of manufacture and erection were enough to invalidate the elastic calculations, which are extraordinarily sensitive to small imperfections of geometry or lack of fit. As has been noted, these conclusions would have caused no surprise to Melan in 1936.

Thus Baker, as Technical Officer of the SSRC, knew that their Final Recommendations for Design were deeply flawed, and, moreover, applied only to the simplest regular array of beams and columns. In a sense, his 1936 text with Pippard is a final exposition of structural analysis as he knew it at that time, offering no possibility of advance or development.

(In fact, the text ran through several editions, and a chapter on plastic methods was added in 1943.)

The elastic designer makes the assumption that the structure is perfect; the calculations, then, refer to that perfect structure, and not to any real construction. The elastic designer's assumptions do, however, seem reasonable, even though they lead to apparently unobservable elastic states; common sense would lead to the belief that a trivial defect cannot really affect the strength of a structure. Common sense is in this instance correct, and the paradox is resolved by concluding that the calculation of elastic stresses is not relevant to the prediction of strength. The strength of a real structure does not depend on an elastic stress reaching some limit at one point in the structure; it is given by the steady development of unacceptably large deformations. It was precisely the study of such behaviour that was reported in the section on plasticity in the 1936 Berlin Congress of the International Association for Bridge and Structural Engineering. Baker went to Germany in the wake of this Congress, and learned from Maier-Leibnitz of the collapse tests on continuous beams.

As has been noted, the collapse loads of such beams are virtually unaffected by practical imperfections of the kind identified by the SSRC. Baker at once set up an intensive investigation of the plastic behaviour of steel structures, first in Bristol in 1936, where he had been appointed professor, and from 1943 in Cambridge. By 1948 the British Standard 449 (The use of structural steel in building) had been altered by the insertion of a clause permitting plastic design.

The work by Baker is summarized in *The Steel Skeleton* volume 2, of 1956. His approach was essentially experimental; he repeated Maier-Leibnitz's tests on continuous beams, and made the first substantial series of tests on portal frames (and later on multi-storey structures). A spectacular application of the theory was to the design of the Morrison shelter, which was installed in over a million households in the UK in World War II. The shelter was designed to absorb the energy of a collapsing house by rotation of plastic hinges, allowing the occupants to remain safe within.

Despite this energy calculation, the plastic analysis of beams and frames was tackled by Baker as a problem in statics; that is, solutions were obtained by drawing bending-moment diagrams. The aim was to represent a *mechanism of collapse* by the superposition of free and reactant bending moments, in the way shown for example by Maier-Leibnitz, fig. 9.2. A very simple example, reproduced from *The Steel Skeleton* volume 2, is shown in fig. 9.4. The free bending moments for the propped

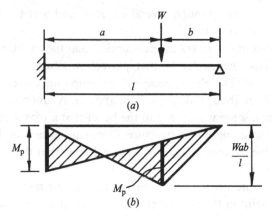

Fig. 9.4. The collapse of a propped cantilever, after *The Steel Skeleton* II.

cantilever are drawn for the equivalent simply supported beam; the maximum bending moment, Wab/ℓ, occurs under the load. The reactant line is positioned to represent the formation of a plastic hinge (full plastic moment M_p) both under the load and at the fixed end, thus creating an elementary mechanism of collapse. It is then a simple problem in geometry to determine the value of M_p as $Wab/(\ell + b)$.

The mechanism of collapse is immediately evident for the beam of fig. 9.4, but it is not so evident for more complex beams and frames. However, the structures that were studied, and the frames designed in practice, were usually simple enough for trial-and-error methods to be adequate. It was of course appreciated that if a structure has n redundancies, then $(n + 1)$ hinges will transform the structure into a 'regular' collapse mechanism. For partial collapse, the number n must be taken to refer to a part only of the structure; many real frames reach the collapse state locally, and become statically determinate there, while the remaining portions remain hyperstatic. Before the late 1940s no principles were generally known that could be applied to the problem of the proper analysis of such cases of partial collapse.

Much of the effort of Baker's teams was therefore directed to other very necessary aspects of practical plastic design, such as the effects of axial load and of shear force on the formation of plastic hinges, which were described in Chapters 2 and 3. Most importantly, elastic/plastic buckling was studied, and this work has been summarized in Chapter 4.

9.7 Gvozdev 1936

The general mathematical principles needed to underpin plastic theory were made available through the activities of W. Prager's group at Brown University. Attempts to establish such principles are evident in the papers of the 1936 Berlin Congress, but the extreme unrest of the times (which led to World War II) put a brake on further theoretical advances. Some of the scholars retired; some of the younger workers left Europe. Prager had not presented a paper in Berlin, but his work was referred to at the 1936 Congress; he left Germany before the outbreak of war, and arrived, via Turkey and India, in the US in 1941. There, at Brown University, he established a Division of Applied Mathematics (with close connexions to the School of Engineering) that attracted outstanding faculty and students. Central to the work of the Division was the advancement of knowledge in the field of plasticity, building on the accumulated knowledge of the 1930s in central Europe.

It turns out that work on plasticity was simultaneously being pursued in the USSR; A.A. Gvozdev presented a paper to a conference on the subject in 1936, which was not however published (in Russian) until 1938. This paper by Gvozdev contains what appear to be the earliest statements and proofs of the fundamental theorems of plasticity (or limit theorems, as they came to be called in the US).

Gvozdev considers an elastic/perfectly plastic material, with a stress–strain (in general, stress resultant–strain resultant) relationship of the type sketched in fig. 9.3; indeed he deals with a rigid/perfectly plastic material, and is precise in his assumption that elastic displacements may be ignored. Specifically, deformation of a structural element can take place only when yield is occurring and, at yield, deformations can increase indefinitely at constant values of the external forces. The set of external forces corresponding to all possible modes of deformation determines the yield condition.

Thus a structural element may be acted upon by a set of generalized forces $s_1, s_2 \ldots s_n$, where these forces might be, for a bending element, bending moments M_x and M_y, axial loads N, and so on. (Equally, the forces $s_1, s_2 \ldots$ could be taken to represent external loads acting on a structure.) A plot may then be made in a space of n dimensions of the yield condition, that is, of those combinations of $s_1, s_2 \ldots s_n$ which will permit unlimited deformations; fig. 9.5(a) shows a two-dimensional plot of this sort. Similarly, a plot may be made in a space of n dimensions where the axes are the corresponding generalized displacement

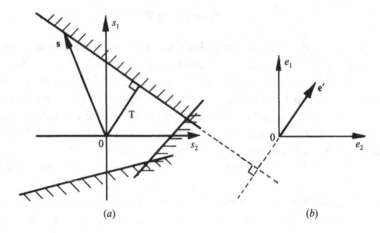

Fig. 9.5. (*a*) Yield surface. The vector **s** represents a combination of forces causing plastic collapse. (*b*) Corresponding strains; the vector **e′** represents a mode of plastic deformation (after Gvozdev).

components $e_1, e_2, \ldots e_n$ of the (unit) vector **e′** which gives the plastic deformation, fig. 9.5(*b*). For a particular displacement vector **e′** the internal work has a certain calculable value, say T, and this must equal the work expended by the force vector **s**, that is

$$T = \mathbf{e}' \cdot \mathbf{s}. \qquad (9.3)$$

Equation (9.3) is represented graphically in fig. 9.5; the ends of all vectors **s**, representing combinations of forces to produce plastic deformation, must lie on a (hyper)plane orthogonal to **e′**. This is the *normality condition* of plasticity theory.

When all possible modes of deformation are considered, a set of hyperplanes is created that define the *yield condition*; three such planes are shown in fig. 9.5(*a*) defining a limiting yield surface. If only discrete modes of deformation are possible, the yield surface will be polygonal; otherwise, the boundaries of the yield surface may be curved. In all cases the yield surface is convex enclosing the origin of the *n*-dimensional space; points within the surface represent combinations of forces for which yield is not occurring.

Gvozdev then proceeds to statements and proofs of the bound theorems of plasticity theory. (The proofs are simply based on the use of virtual work, and are very similar to those given below.) The loading on the structure is taken to be proportional, so that all loads may be specified in terms of one of their number, P say. Gvozdev specifies the three conditions that must be satisfied at collapse, and these may be labelled

$$\left.\begin{array}{ll} \text{Equilibrium condition} \\ \text{Yield condition} \end{array}\right\} \text{ I (Statically admissible)} \left.\begin{array}{l} \\ \\ \\ \end{array}\right\} \quad (9.4)$$
$$\begin{array}{ll} \text{Mechanism condition} & \text{II (Kinematically admissible).} \end{array}$$

(The bracketed terms 'statically admissible' and 'kinematically admissible' were introduced by Greenberg and Prager (1952).) The problem is to determine the value P_c of the collapse load at which all three of the conditions are satisfied.

Gvozdev first considers states I of the structure for which the equilibrium and yield conditions are satisfied, say at a load P_I. Evidently there will be a largest value P_I^{max}. If now a displacement of the structure is considered that is caused solely by deformations of the plastic zones, then a value of load P_{II} may be calculated by equating the work done by P_{II} on the displacements of the structure to the internal plastic work. When all possible deformations are considered, then a smallest value P_{II}^{min} may be determined. Since at collapse all three conditions must be satisfied, then

$$P_I^{max} \geq P_c \geq P_{II}^{min}. \tag{9.5}$$

A simple *reductio ad absurdum* argument, given below, shows that

$$P_I^{max} = P_{II}^{min}, \tag{9.6}$$

from which the basic theorems follow:

- A: *Uniqueness.* The collapse load has a definite value $P_c = P_I^{max} = P_{II}^{min}$.
- B: *Upper bound.* A value P_{II} calculated from a possible mode of plastic displacement is an upper bound on the value of the collapse load, that is, $P_{II} \geq P_c$.
- C: *Lower bound.* A value P_I calculated from an equilibrium state which satisfies the yield condition is a lower bound on the value of the collapse load, that is, $P_I \leq P_c$.

9.8 Proofs of the plastic theorems

These fundamental results obtained in 1936 by Gvozdev were unknown outside Russia, and little noticed within that country. The static (or lower-bound, or safe) theorem was restated in 1948 in Russian by Feinberg, but without proof, and came to the attention of Prager. Greenberg and Prager supplied proofs in 1949 of the upper and lower bound theorems (published 1952), and Horne (1950) added a proof of the uniqueness theorem. These mathematical theorems provided the rigorous backing for Baker's advances in engineering plasticity; Baker and Prager met in 1947, and there followed fruitful collaboration between Cambridge and Brown Universities.

The proofs of the theorems follow, as has been indicated, from straight-forward applications of the equation of virtual work, and can perhaps be given most easily in terms of the simple framed structure. (The theorems hold, subject to the usual assumptions of small displacements and stable behaviour, for more general continuum structures.) Figure 9.6(a) shows a rectangular portal frame acted upon by loads W; all loads are acted upon by the same multiplying load factor λ. The value λ_c at collapse is sought. In terms of this frame Gvozdev's three conditions (9.4) may be written as:

Equilibrium:	Internal bending moments M in the frame are in equilibrium with the external loads W.
Yield:	The values of M are less than, or at most equal to, the value of the full plastic moment M_p.
Mechanism:	There is an arrangement of plastic hinges which will permit deformation of the frame.

Figure 9.6(b) shows a mechanism of deformation, and the rotations θ are compatible with displacements δ of the loading points. The internal work dissipated at a plastic hinge is $M_p|\theta|$, which is always positive; the value of M_p may vary from point to point round the frame.

Uniqueness. It will be supposed that, for a given loading on a frame, there are two different collapse mechanisms formed at different load factors λ^* and λ^{**}. For the first mechanism the collapse bending moments round the frame are given by a distribution M^*, where the equilibrium equations are satisfied and $|M^*| \leq M_p$; the mechanism of collapse is (δ^*, θ^*). A similar statement may be made for collapse at the load factor λ^{**}, so that

$$
\left.
\begin{array}{ll}
\text{A:} & (\lambda^* W, M^*) \text{ satisfy the equilibrium and yield conditions} \\
\text{B:} & (\lambda^{**} W, M^{**}) \text{ satisfy the equilibrium and yield conditions} \\
\text{C:} & (\delta^*, \theta^*) \text{ describes a mode of plastic deformation} \\
\text{D:} & (\delta^{**}, \theta^{**}) \text{ describes a mode of plastic deformation.}
\end{array}
\right\}
\quad (9.7)
$$

The collapse equation for the first mechanism may be written by combining statements A and C in the equation of virtual work:

$$\Sigma \lambda^* W \delta^* = \Sigma M^* \theta^*. \tag{9.8}$$

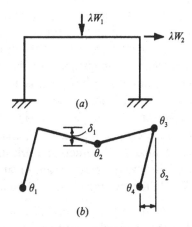

Fig. 9.6. (a) A typical frame acted upon by loads W. (b) A typical mode of plastic deformation.

The value of $|M^*|$ at each hinge position is equal to M_p, so that the collapse load factor λ^* is given by

$$\lambda^* \Sigma W \delta^* = \Sigma M_p |\theta^*|. \tag{9.9}$$

Statements B and C in (9.7) can also be combined by the equation of virtual work:

$$\lambda^{**} \Sigma W \delta^* = \Sigma M^{**} \theta^*. \tag{9.10}$$

The bending moments M^{**} satisfy the yield condition; that is, if mechanisms θ^* and θ^{**} have a common hinge, then $|M^{**}| = M_p$ at that hinge, but otherwise $|M^{**}| < M_p$ at the hinge points of the mechanism θ^*. Thus, in equation (9.10),

$$\Sigma M^{**} \theta^* \leq \Sigma M_p |\theta^*|, \tag{9.11}$$

so that

$$\lambda^{**} \Sigma W \delta^* \leq \Sigma M_p |\theta^*|. \tag{9.12}$$

Comparing equation (9.9) with inequality (9.12),

$$\lambda^{**} \leq \lambda^*. \tag{9.13}$$

Statement D in (9.7) has not been used, and if the arguments are repeated with statement D instead of statement C,

$$\lambda^* \leq \lambda^{**}. \tag{9.14}$$

Thus λ^* and λ^{**} have the same value, namely the collapse value λ_c. This, with a different notation, is essentially the proof given by Gvozdev for a generalized plastic body. It may be noted that the proof has shown only that the load factor at collapse is unique. Nothing has been proved about the mode of deformation, and indeed it is possible for different modes to exist at the same value of collapse load factor.

The upper bound theorem (the unsafe theorem). The theorem states that if a plastic mode of deformation is assumed, and the work done by the external loads is equated to the internal work dissipated, then the resulting load factor λ' is always greater than, or at best equal to, the true load factor λ_c. The following statements will be used:

$$\left. \begin{array}{ll} \text{E:} & (\lambda_c W, M_c) \text{ is the actual collapse distribution} \\ \text{F:} & (\delta', \theta') \text{ is the assumed collapse mechanism.} \end{array} \right\} \qquad (9.15)$$

The work equation for the assumed collapse mechanism is

$$\lambda' \Sigma W \delta' = \Sigma M_p |\theta'|. \qquad (9.16)$$

Statements E and F of (9.15) combine to give

$$\lambda_c \Sigma W \delta' = \Sigma M_c \theta'. \qquad (9.17)$$

Now $|M_c| \leq M_p$, so that, following the previous arguments,

$$\lambda_c \Sigma W \delta' \leq M_p |\theta'|, \qquad (9.18)$$

and comparison of (9.16) and (9.18) shows that

$$\lambda_c \leq \lambda'. \qquad (9.19)$$

The lower bound theorem (the safe theorem). The theorem states that if a set of bending moments can be found that satisfies the equilibrium and yield conditions at a load factor λ'', then λ'' is always less than, or at best equal to, the true load factor λ_c. The following statements will be used:

$$\left. \begin{array}{ll} \text{G:} & (\lambda'' W, M'') \text{ represents a set of bending moments} \\ & \text{satisfying the equilibrium and yield conditions} \\ \text{H:} & (\lambda_c W, M_c) \text{ is the actual collapse distribution} \\ \text{J:} & (\delta_c, \theta_c) \text{ is the actual collapse mechanism.} \end{array} \right\} \qquad (9.20)$$

Statements H and J give

$$\lambda_c \Sigma W \delta_c = \Sigma M_c \theta_c = \Sigma M_p |\theta_c|, \qquad (9.21)$$

while G and J give

$$\lambda'' \Sigma W \delta_c = \Sigma M'' \theta_c \leq \Sigma M_p |\theta_c| \qquad (9.22)$$

as before. Hence

$$\lambda'' \leq \lambda_c. \tag{9.23}$$

Thus Gvozdev's three conditions (9.4) that must be satisfied at collapse of a plastic body may be displayed compactly as the three theorems:

$$\lambda = \lambda_c \left\{ \begin{array}{l} \text{Equilibrium condition} \\ \text{Yield condition} \\ \text{Mechanism condition} \end{array} \right\} \left. \begin{array}{l} \lambda \leq \lambda_c \\ \lambda \geq \lambda_c \end{array} \right\} \tag{9.24}$$

9.9 Methods of calculation

With fundamental theorems available from about 1950, the way was open for the development of techniques of calculation. These were devised first for use by hand, but they coincided with the wide-scale development of electronic computing, and rapid progress was made in methods of plastic design of framed structures. These are now part of the general vocabulary of standard texts, and Neal and Symonds are responsible for advances in at least two areas. One of these, the method of *combination of mechanisms*, powerful as a hand method but easily programmable, is based on the fact that very few independent equilibrium equations exist for a framed structure. Indeed, if a frame has a number R of statical indeterminacies (the frame in fig. 9.7(a) has 18 redundancies), and there are N possible critical sections at which plastic hinges might form (36 in fig. 9.7), then there exist $(N - R)$ equilibrium equations connecting the values of the bending moments at the critical sections with the magnitudes of the applied loads.

Neal and Symonds (1952) made use of the fact that an equilibrium equation can be generated, through the use of virtual work, from a corresponding mechanism. There is thus a one-to-one correspondence between an independent equilibrium equation and an independent mechanism; just as all possible statements of equilibrium for a frame can be deduced from the independent equilibrium equations, so all possible mechanisms of collapse can be built up from the independent elementary mechanisms. Thus $(N - R)$ has the value 18 for the frame of fig. 9.7, and these 18 mechanisms comprise 6 of elementary beam type, of which one is sketched in fig. 9.7(b), 3 sways, one of each storey, of which one is sketched in fig. 9.7(c), and 9 degenerate joint mechanisms (expressing the fact that the moments acting on the ends of members meeting at a joint

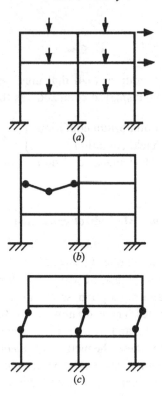

Fig. 9.7. (*a*) Multi-storey multi-bay frame. (*b*) Typical elementary beam mechanism. (*c*) Typical storey sway.

must sum to zero). From these simple elements can be built up a highly complex final collapse mechanism.

It will be seen from statements (9.24) above that the method is 'unsafe'; mechanisms are examined, and the objective is to reduce to its minimum the value of the loading parameter. The simplicity of the method derives from the fact that only one of the three collapse conditions is examined; no attempt is made to satisfy (during the course of the calculation) the equilibrium or yield conditions. As a trivial example, the rectangular portal frame of fig. 9.8 has members of uniform full plastic moment M_p; it is required to find the collapse value λ_c of the load factor λ. The frame has 5 critical sections and 3 redundancies; the two independent mechanisms may be taken as those sketched in fig. 9.8(*b*) and (*c*), and

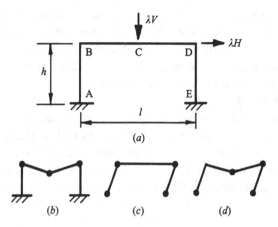

Fig. 9.8. (a) A simple rectangular portal frame. (b) and (c) Two independent modes of deformation. They combine to give mode (d).

give the collapse equations

$$\left.\begin{aligned}\lambda\frac{V\ell}{2} &= 4M_{\mathrm p}\\ \text{and}\quad \lambda Hh &= 4M_{\mathrm p}.\end{aligned}\right\} \tag{9.25}$$

Only one combination is possible; fig. 9.8(d) results from the super-position of (b) and (c), eliminating the hinge at corner B, and the corresponding collapse equation is

$$\lambda\left(\frac{V\ell}{2} + Hh\right) = 6M_{\mathrm p}. \tag{9.26}$$

From the upper-bound theorem, that mechanism is correct (for given loads and dimensions of the frame) which gives the smallest value of λ from the three equations (9.25) and (9.26).

More generally, if the bending moment at section i ($i = A,\ldots,E$) has value M_i, then (with an appropriate sign convention) the two independent mechanisms (b) and (c) of fig. 9.8 give the equilibrium equations

$$\left.\begin{aligned}M_B - 2M_C + M_D &= \lambda\frac{V\ell}{2}\\ \text{and}\quad M_A - M_B \quad + M_D - M_E &= \lambda Hh.\end{aligned}\right\} \tag{9.27}$$

To these equations may be added the yield condition

$$-M_{\mathrm p} \le M_i \le M_{\mathrm p}. \tag{9.28}$$

It will be seen that the first of equations (9.25), for example, may be derived from the first of (9.27) by maximizing, numerically, the value of the left-hand side; that is, the value of λ from equations (9.27), and from all other possible equilibrium equations, is made as *large* as possible. The *smallest* value of λ from all these possibilities is then the correct value, λ_c. This 'minimax' behaviour was, as has been seen, remarked on by Coulomb (1773).

In fact, Neal and Symonds (1950–51) had proposed earlier a solution of the plastic problem working from equations such as (9.27) and (9.28), that is, based on the satisfaction of the equilibrium and yield conditions. This is a 'safe' approach, and the problem posed in this way is a standard one of linear programming; Charnes and Greenberg (1951) identified this application.

9.10 Shakedown and incremental collapse

According to Neal (1956), both Grüning (1926) and Kazinczy (1931) recognized that a structure might fail under repeated loading by excessive plastic flow, even though no single load combination were severe enough to cause plastic collapse. Inequalities (9.1), as stated by Bleich, may be written

$$\left.\begin{array}{l} \lambda \mathscr{M}^{\max} + m \le M_{\mathrm{p}} \\ \lambda \mathscr{M}^{\min} + m \ge -M_{\mathrm{p}}, \end{array}\right\} \tag{9.29}$$

where the device of a proportional load factor λ has been introduced. A conventional elastic solution is obtained for the various specified loads acting on a frame, and the bending moments $\mathscr{M}^{\max}$ and $\mathscr{M}^{\min}$ are calculated at each cross-section to give the greatest and least values; these values are then multiplied by the factor λ.

The shakedown (lower-bound) theorem states that if a set of self-stressing moments m can be found so that inequalities (9.29) are satisfied, then the structure will shake down at a load factor λ. That is, although some plastic deformation may occur under initial applications of the applied loading, the structure will eventually resist further loading solely by responding elastically. Inequalities (9.29) are clearly necessary if plastic flow is not to occur; the shakedown theorem states that they are also sufficient.

From inequalities (9.29),

$$M_{\mathrm{p}} - \lambda \mathscr{M}^{\max} \ge m \ge -M_{\mathrm{p}} - \lambda \mathscr{M}^{\min}, \tag{9.30}$$

$$\text{that is,} \quad \lambda \left(\mathscr{M}^{\max} - \mathscr{M}^{\min} \right) \le 2M_{\mathrm{p}}. \tag{9.31}$$

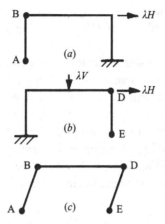

Fig. 9.9. Development of an incremental collapse mechanism. (*a*) Side load causes small rotations of hinges A and B. (*b*) Side and vertical loads cause small rotations of hinges D and E. (*c*) State of frame after a few cycles of loading.

Inequality (9.31) describes the first way in which plastic failure might occur under repeated loading; if it is not satisfied at a particular cross-section, that section would be bent back and forth, yielding in first one sense and then the other, with the possibility of low-cycle fatigue failure. For normal loading on usual civil-engineering structures, such alternating plasticity is unlikely, although it is certainly possible in constructions such as pressure vessels.

The second way in which plastic failure might occur is by incremental collapse. At some stage in the loading cycle, equality is reached in one of the inequalities at one or more critical sections of the frame, so that a plastic hinge is formed and plastic rotation occurs. There are not, however, sufficient hinges to turn the whole frame into a mechanism of collapse, so that plastic deformation is constrained to be small by those portions of the frame remaining elastic. Under a different loading combination other hinges might form, and again irreversible plastic rotation could occur. Thus in fig. 9.9 the side load λH might produce plastic hinges at A and B, whereas on some other occasion loads λH and λV might produce hinges at D and E. Neither fig. 9.9(*a*) nor (*b*) represents a mechanism of collapse, but after a few repetitions of such cyclic loading the incremental collapse mechanism of fig. 9.9(*c*) could be well developed.

Melan (1936) gave a proof that inequalities (9.29) were sufficient to prevent such incremental collapse – his proof was for pin-jointed trusses

(assuming stable behaviour in compression). The proof was later simplified by Symonds and Prager (1950), and Neal (1951) adapted it for
frames, as follows.

A small change is considered in the values of the applied loads. This
variation of the loads may cause some yielding to occur, so that not only
will the elastic moment change to $(\mathcal{M}_i + \delta\mathcal{M}_i)$, but the residual moment
m_i at the same section i may change to $(m_i + \delta m_i)$. During this process
the change of curvature $(\delta\mathcal{M}_i + \delta m_i)/EI$ at each section of the frame
that remains elastic will be compatible with any hinge rotations $\delta\theta_k$ that
may occur at the yielding sections k. Thus, since m_i is a set of bending
moments in equilibrium with zero external load, the equation of virtual
work gives

$$\int m_i \frac{\delta\mathcal{M}_i}{EI} ds + \int m_i \frac{\delta m_i}{EI} ds + \Sigma m_k \delta\theta_k = 0, \qquad (9.32)$$

where the integration extends over all portions of the frame that remain
elastic during the change in applied loading, and the summation includes
all hinge rotations that occur.

The moments $\mathcal{M}_i$ are bending moments computed for a frame that is
wholly elastic, so that the changes of curvature $\delta\mathcal{M}_i/EI$ give a compatible
set of deformations of the frame; a second and separate application of
virtual work shows therefore that the first integral in equation (9.32) must
be zero.

A set of self-stressing bending moments satisfying inequalities (9.29)
will be denoted $\bar{m}_i$. Exactly the same arguments then show that

$$\int \bar{m}_i \frac{\delta m_i}{EI} ds + \Sigma \bar{m}_k \delta\theta_k = 0, \qquad (9.33)$$

and equations (9.32) and (9.33) combine to give

$$\int (m_i - \bar{m}_i) \frac{\delta m_i}{EI} ds + \Sigma (m_k - \bar{m}_k) \delta\theta_k = 0. \qquad (9.34)$$

It is now supposed that at a particular section k where yield is occurring,
the current value of m_k is such that $m_k < \bar{m}_k$, that is

$$(m_k - \bar{m}_k) < 0. \qquad (9.35)$$

The first of inequalities (9.29) gives

$$\lambda\mathcal{M}_k^{\max} + \bar{m}_k \le \left(M_p\right)_k, \qquad (9.36)$$

$$\text{that is} \qquad \lambda\mathcal{M}_k^{\max} + m_k < \left(M_p\right)_k, \qquad (9.37)$$

where, in inequality (9.37), the possibility of equality no longer exists.

Since, however, it was supposed that yield *is* occurring at the section, it must be under the negative value $-(M_\mathrm{p})_k$ (i.e. the second inequality (9.29) must just be satisfied), and the corresponding value $\delta\theta_k$ of the hinge rotation must also be negative. Thus, using (9.35),

$$(m_k - \overline{m}_k)\,\delta\theta_k > 0. \tag{9.38}$$

In the same way, if it is assumed that $m_k > \overline{m}_k$, then of necessity $\delta\theta_k > 0$. If $m_k = \overline{m}_k$, the sign of $\delta\theta_k$ cannot be determined, but in all cases

$$(m_k - \overline{m}_k)\,\delta\theta_k \geq 0. \tag{9.39}$$

Hence from equation (9.34)

$$\int (m_i - \overline{m}_i)\,\frac{\delta m_i}{EI}\,ds \leq 0. \tag{9.40}$$

Now the quantity

$$U = \int \frac{(m_i - \overline{m}_i)^2}{2EI}\,ds \tag{9.41}$$

is positive definite, and inequality (9.40) states that, as the loading on the frame changes, $\delta U \leq 0$. Thus the value of U can only decrease if any plastic deformation occurs, and it remains constant otherwise; the value of U either must become zero, in which case $m_i = \overline{m}_i$ everywhere, or must settle down to a definite positive value. All further changes in the applied loading will then be resisted purely elastically, and the frame will have shaken down.

As for the case of fixed static loading, simpler calculations result for variable loading if attention is concentrated on possible incremental modes of plastic deformation, rather than on solutions satisfying the yield condition, inequalities (9.29). A particular mechanism is assumed, say that of fig. 9.9(c), where the hinges are actually formed at different stages in the loading cycle, figs 9.9(a) and (b). If the hinge rotation θ_i at one of the hinges is positive, say $+\theta_i^+$, then a hinge will form when

$$\lambda \mathscr{M}_i^{\max} + m_i = (M_\mathrm{p})_i, \tag{9.42}$$

and if the rotation is negative, say $-\theta_i^-$, then

$$\lambda \mathscr{M}_i^{\min} + m_i = -(M_\mathrm{p})_i. \tag{9.43}$$

If equation (9.42) is multiplied through by $+\theta_i^+$, and equation (9.43) by $-\theta_i^-$, then either

$$\left.\begin{array}{r} \lambda \mathscr{M}_i^{\max}\theta_i^+ + m_i\theta_i = (M_\mathrm{p})_i\,|\theta_i| \\ \text{or} \quad -\lambda \mathscr{M}_i^{\min}\theta_i^- + m_i\theta_i = (M_\mathrm{p})_i\,|\theta_i|. \end{array}\right\} \tag{9.44}$$

Equations (9.44) can now be summed for all hinges of the assumed incremental collapse mechanism, and since, as usual, $\Sigma m_i \theta_i = 0$, then

$$\lambda \left(\Sigma \mathscr{M}_i^{\max} \theta_i^+ - \Sigma \mathscr{M}_i^{\min} \theta_i^- \right) = \Sigma \left(M_p \right)_i |\theta_i|. \qquad (9.45)$$

This is the incremental collapse equation and, using arguments similar to those involved in the proof of the upper-bound theorem for static loading, it is easily shown that the value of λ resulting from equation (9.45) is an unsafe estimate of the true incremental collapse (or shakedown) factor λ_s, i.e.

$$\lambda \geq \lambda_s. \qquad (9.46)$$

It may be noted that for *fixed* loading leading to static collapse at a load factor λ_c, equation (9.45) gives

$$\lambda_c \Sigma \mathscr{M}_i \theta_i = \Sigma \left(M_p \right)_i |\theta_i|; \qquad (9.47)$$

the static plastic collapse load factor for a given mechanism θ may be calculated immediately from the elastic solution $\mathscr{M}$.

It seems apparent, and indeed is so, that the value of λ_s calculated from the incremental collapse equation (9.45) cannot exceed the value of λ_c calculated from the static equation (9.47), in which all loads are given their fixed maximum values. That is, variable loading is always more critical for a structure than fixed loading of the same magnitudes. This matter was discussed by Ogle (1964); see also Heyman (1971).

9.11 Coda

It is, of course, not necessary to obtain first the elastic solution in order to make a collapse analysis for a structure, as might be implied by equation (9.47). Historically this was in fact the approach, as may be seen in the work of Maier-Leibnitz and Bleich reported to the 1936 Berlin Congress. However, equation (9.22) may be rewritten

$$\Sigma M_W \theta = \Sigma M_p |\theta|, \qquad (9.48)$$

and this may be regarded as the basic equation for making a plastic design of a frame. Here M_W represents *any* set of bending moments in equilibrium with the (factored) loads acting on the structure; the external loading has been replaced by this equilibrium set M_W. Since equation (9.48) represents an 'unsafe' design for an arbitrarily assumed mechanism of deformation θ, all such mechanisms must be examined, and that one is correct which gives the largest value of M_p.

The elastic bending moments $\mathcal{M}$ are, naturally, in equilibrium with the external loads; equation (9.47) is merely a special, but very interesting, case of equation (9.48).

It will have been noted that, in the proofs of the theorems, sets of self-stressing moments are introduced which are superimposed on the moments (M_W, say, as in equation (9.48) above) that equilibrate the applied loads. These self-stressing moments could be induced in the structure by plastic yield on first loading, or by the forcing together of members during manufacture to rectify imperfect geometry, or by settlement of supports, or by temperature strains, or by the firm bolting together of members at a connexion that is regarded as pinned (secondary stresses). However they arise, states of self-stress, by definition in equilibrium with zero external load on the structure, cannot affect the simple plastic theorems. The collapse load for a ductile structure whose members do not become unstable is unique, and is independent of any initial or induced state of self-stress.

10
Scholium 2

Navier (1826) identified the three, and only three, groups of equations that can be formulated to analyse a structure. Foremost are the equations of equilibrium, which relate the internal forces to the given externally applied loads. If these equations alone determine the internal forces, then the structure is, by definition, statically determinate.

In general, a structure is hyperstatic, and the other two sets of equations must be used in order to solve the prime structural problem, that of finding the internal forces. Statements must be made about how the internal forces are related to internal deformations – a 'stress–strain' relationship must be specified, and, until the advent of plastic methods, this relationship was usually taken to be linear-elastic. Other material properties may also come into play in calculating the internal deformations – for example, strains due to temperature. Finally, the equations of compatibility are used to make geometrical statements; the members are constrained to fit together, internal deformations must be related to external movements of the structure, and the structure as a whole is constrained by its attachment to its environment.

10.1 Hambly's paradox

Hambly (1985) posed a pedagogic problem to illustrate the difficulties of design of a hyperstatic structure:

A milkmaid weighing 600 N sits on a three-legged stool. For what basic force should each leg of the stool be designed?

The stool is supposed to be symmetrical, the milkmaid sits at the centre of the seat, and so on. The answer to the question is, of course, 200 N.

The same milkmaid now sits on a square stool with four legs, one at each corner, and again the stool and the loading are symmetrical. For what force should each leg of the stool be designed? The answer of 150 N is not necessarily correct. A robust, nearly rigid milking stool, standing on a firm, nearly rigid milking-shed floor, will rock; three of the legs will appear to be in contact, supporting the weight of the milkmaid, but the fourth will be clear of the floor. If this fourth leg is clear by only a fraction of a millimetre, then it is certain that the force it is carrying is zero. By simple statics, the force in the leg diagonally opposite will also be zero, even if it seems to be touching the floor. The weight of the milkmaid is in fact supported symmetrically by the other two legs of the stool, and each must therefore be designed to carry a force of 300 N.

Now the stool may be imagined to be placed on a randomly rough floor, and there is no way of deciding *a priori* which legs are in contact – *all* legs must therefore be designed to carry a force of 300 N. This is the paradox – the addition of a fourth leg implies an increase, rather than a decrease, in the force for which each leg must be designed.

10.2 The elastic design process

Only three equations of equilibrium are available for the four-legged stool, and there are four leg forces to be found. For this simple problem the leg forces are equal to the supporting reactions from the floor, R_1, R_2, R_3 and R_4 say. If the milkmaid sits centrally but no assumption is made about any consequent symmetry in the values of the reactions, then at least it is known that their sum must be 600 N. The two further equations, obtained by taking moments, show that diagonally situated legs carry equal loads, i.e. $R_1 = R_3$ and $R_2 = R_4$. Thus $R_1 + R_2 = 300$, but no further information results from the equilibrium equations. However, the physical problem requires that both R_1 and R_2 be positive, so that $0 \leq R_1, R_2 \leq 300$.

(It may be remarked that the statement that only three equations of overall equilibrium are available for analysis of the stool results from a preliminary simplification in the modelling of the problem. For example, the designer will have assumed that the legs make point contact with the floor – the leg-ends are rounded, perhaps. Further, it will have been assumed that the reactions on the legs from the floor are vertical – the floor is smooth. If in fact contact is rough, then horizontal forces may act at the feet of the legs; extra equations of equilibrium may be written, but the degree of structural indeterminacy rises sharply. Such considerations may or may not be of importance in the final assessment of the design requirements for a leg of the stool.)

The four-legged stool, then, cannot be solved by statics alone, and Navier showed the way to proceed. Elastic information must be introduced – the flexural properties of the slab forming the seat of the stool must be specified, as must the axial compressibility of the legs. The analysis at once becomes complex, since the flexure of a square slab supported at four corners is not a simple problem. In fact, the designer might well wish to make the common-sense assumptions that the slab is rigid and the legs are incompressible, but no solution would then be possible – an elastic analysis requires a knowledge of elastic constants. A compromise would be to assume a rigid slab, but to allow elastic compressibility of the legs. A straightforward analysis then results; no matter what value of the elastic constant is taken for the legs, the force in each leg is found to be 150 N.

This result has been obtained, of course, on the assumption, without consideration, of uniform boundary conditions; the floor is rigid and level, and all legs of the stool are of the same initial length so that they make the same contact with the floor. In fact, as has been noted, the boundary conditions are unknown and unknowable – the milkmaid will place the stool randomly on a rough floor. It may well be concluded that trivial irregularities can really have no significant influence on the problem of designing a safe stool, but elastic analysis does not support this common-sense view. If the stool is analysed with one leg clear of the floor, then profoundly different values for the elastic leg forces will be found for clearances of say 0.01, 0.1 and 1 mm. These different values will be confirmed by experiment; if tests are carried out, with gauges attached to the legs, then the load in a leg may be found to have any value lying between 0 and 300 N, and a fair number of experiments will record the load as exactly 0 or 300 N.

It was precisely observations of this sort that were made by the Steel Structures Research Committee in the 1930s, and their conclusion was that the host of geometrical imperfections in structures made elastic analysis the wrong tool for design. It was these observations, coupled with the experimental work of Kazinczy, Maier-Leibnitz and others, that led to the plastic method of structural design for steel structures (or indeed of a structure made of any ductile material).

10.3 Simple plastic design

The essence of the words 'plastic' and 'ductile' is that a limiting stable behaviour is implied. For the simple example of a transversely loaded steel beam, an imagined slow increase in load will lead to the development of a plastic hinge. Rotation can take place at this hinge leading to large structural deflexions, but the process is quasi-static, with the applied loads remaining constant. Reinforced concrete shows this same kind of behaviour, at least for the range of displacements experienced by usual structures. In fact any material (aluminium alloy, wood) used in practical design is ductile in this way, as opposed to brittle materials (glass, cast iron) which will crack if overstrained and could lead to catastrophic collapse. Galileo's marble column fractured when stored on its side in an accidentally hyperstatic situation.

If, then, a plastic analysis is to be made of the stool, it must be ensured that the final failure of a leg is ductile. This implies that each leg should be 'stocky', and capable of sustaining a limiting 'squash load' without exhibiting unstable behaviour. The loading history of such a stool may now be followed (as might have been done in the mid-1930s) as a point load at the centre is slowly increased.

Initially, in general, only two legs will carry the load. If the squash load of each leg is P, then these two legs will yield, and compress stably, when the point load has value $2P$. The compression of the yielded legs will allow the unloaded legs to make contact with the floor, and the point load may then be increased above the value $2P$ – in fact, to the value $4P$. At this final stage all four legs are carrying their maximum loads, and a large displacement of the seat of the stool will follow. The plastic design of the stool to carry a milkmaid of 600 N would therefore require the value of P to be 150 N, and this is the force for which each leg should be designed – of course, with a suitable factor. (It may be that a load factor, say 3, has already been incorporated; that is, the stool is actually being designed for the sole use of the milkmaid's daughter, who weighs 200 N, and the milkmaid herself is a fiction. Alternatively, Bleich's approach may be adopted; the milkmaid is real, and weighs 600 N, so that each leg of the stool will be manufactured with a strength three times that calculated, that is, with a squash load of 450 N.)

The plastic solution – each leg to be designed for 150 N – has been obtained by following the particular loading history outlined above, but other histories are possible. For example, one leg may be initially only just clear of the floor, and application of the load may bring this leg into contact through *elastic* shortening of the two loaded legs. However, further increase of loading will eventually, in the symmetrical case, cause two legs to yield, and the load can be increased until collapse occurs, once more at a value of $4P$. Or again, the stool may be positively pinned to the rough floor, so that the legs can take both tensile and compressive forces; initial installation will induce a state of self-stress, in which two opposite legs carry the same value of tensile force, while the other two carry the same force in compression. Application and slow increase of the central load will cause the two legs in compression to yield first; the final collapse load of $4P$ remains the same.

Simple plastic design, directed to the calculation of ultimate loads, is independent of the loading path, of initial imperfections, of temperature strains and of initial states of self-stress.

10.4 Design with unstable elements

Hambly's paradox does not exist either for the straightforward elastic designer or for the simple plastic designer – they are united in the view that the legs of the four-legged stool should be designed for 150 N to support a weight of 600 N. However, plastic design, and its comforting 'safe' theorem, is valid only if the structural elements (the legs) are stable. Should a leg buckle, the load/deflexion characteristic for that leg will peak and fall, rather than display the required 'ductility' at constant failure load.

For example, the simple plastic designer, aware of the buckling danger but relying mistakenly on the safe theorem, may design each leg to buckle at 160 N. If the stool is placed carefully and exactly on a flat floor, then a 600 N load will induce leg forces of 150 N, and all will be well. On a rough floor, however, one leg will not be in initial contact, and the (factored) load will cause (at least) one leg to buckle and to deform permanently, leading to collapse of the stool. Put another way, the load factor against collapse has been reduced substantially to 1.6 from the design value of 3.

These arguments apply in exactly the same way to the elastic designer, who is also designing against instability and who also believes the forces in the legs all to be 150 N. The fact is that elastic analysis cannot be used since the boundary conditions are unknowable, and the resulting design may be unsafe; equally, plastic design is invalid, and unsafe, if there are unstable elements in the structure. Hambly's question appears not to have been answered: for what force should each leg of the stool have been designed?

The question has in fact been addressed, and a partial answer may be found, for example in the 'weak-beam strong-column' philosophy of design proposed by a Joint Committee (1964) for the design of multi-storey frames. The beams in such a structure are taken to be stable elements (which may involve restraint against lateral buckling), so that they can be designed to have some sort of minimum section – that is, a plastic design would be appropriate. The columns of the multi-storey frame, however, are given cross-sections that may be more than adequate from the point of view of strength, but ensure that unstable buckling cannot occur.

It is the word 'ensure' that causes difficulties. If the end conditions on a column are known, then, despite the fact that buckling is sensitive to imperfections, to residual stresses and so on, the rational/empirical methods outlined in §4.6 will give assurance of stability. In any case of doubt, a column can be replaced by one slightly larger, and certainly stable, at almost no penalty of cost or weight. It is, however, the determination of the end conditions that is not easy. Neither the elastic designer, working with a conventional design procedure, nor the plastic designer, using known values of bending moments applied to the columns from collapsing beams, may be aware of the problem, but the problem nevertheless exists.

An illustrative example is shown in fig. 10.1. A simple rectangular portal frame of uniform section is designed plastically as in fig. 10.1(a); the full plastic moment M_p is of value 6 units. Figure 10.1(b) shows the corresponding collapse analysis; the frame has one redundancy left at collapse, and the bending moments shown satisfy the equilibrium equations. The yield condition will also be satisfied if the unknown bending moment M has any value between 6 and -1.2; under these conditions the solution is 'safe', and the collapse analysis is confirmed provided that neither column becomes unstable.

It is clear that $M = -1.2$ will give the worst bending-moment distribution for design of the columns, fig. 10.1(c); both columns are bent in single curvature. Similarly, $M = 6$ will give the most favourable distribution, fig. 10.1(d). A full elastic–plastic analysis, with the assumptions of perfect fixity of the column feet, gives the bending moments of fig. 10.1(e), which resembles the distribution corresponding to a fully elastic analysis, fig. 10.1(f). The distribution of fig. 10.1(c) should of course be used to ensure that the columns remain stable, but the whole matter is largely unaddressed by any current design process. The straightforward elastic designer will accept fig. 10.1(f); the plastic designer is forced to make some decisions, since the simple analysis does not give the value of the unknown bending moment M. The elastic–plastic analysis, fig. 1.10(e), may well be used. It may be noted that a small settlement of the left-hand

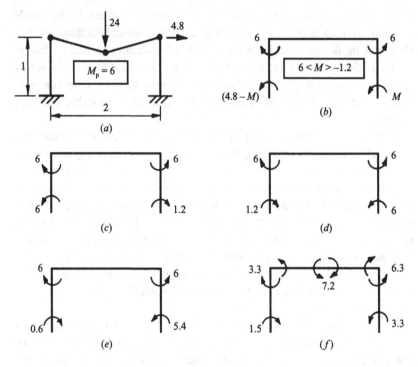

Fig. 10.1. (a) Plastic design of a uniform portal frame. Full plastic moment M_p is 6 units at unit load factor. (b) Analysis at collapse. One redundancy remains: M can have any value between 6 and -1.2. (c) Worst condition for stability of columns, corresponding to slight settlement of left-hand column. (d) Similarly, best conditions for columns. (e) Column moments given by elastic–plastic analysis, assuming perfect fixity at column feet. (f) Similarly, wholly elastic analysis.

footing will modify the elastic–plastic distribution to that of the most severe case, fig. 10.1(c).

Thus, to revert to the problem of the four-legged stool, it has been seen that a simple plastic design leads to forces of 150 N in the legs under a load of 300 N. If the legs are ductile and stable, then they may be designed safely for a load of 150 N; if this assurance of stability cannot be given, they must be designed for 300 N.

10.5 Conclusion

The theory of structures is concerned with the mechanics of slightly deformable bodies. If a structure is to be of practical use, its displacements must be very small; the structure as a whole, and its components, are almost, but not quite, rigid, so that very small deformations engender large internal forces. These internal forces must satisfy the equations of equilibrium that connect their values with those of the external loads, but, for a hyperstatic structure, there exist an infinite number of solutions of the equilibrium equations. That solution is correct which satisfies also the geometrical boundary conditions.

This is the 'Navier' statement of the structural problem, and it leads to a solution which cannot be observed in practice, as was shown explicitly by the Steel Structures Research Committee in the 1930s, and implicitly at about the same time by those developing plastic methods. Of the three sets of master equations, those of equilibrium must certainly be satisfied. The material properties, such as Young's modulus and the yield stress, used in the second set of equations, are known accurately for steel, and fairly well for reinforced concrete, timber and other structural materials. These material properties are needed in order to calculate deformations, so that the third set of equations may be written, expressing geometrically the compatibility of deformation. It is these compatibility equations which are either poorly known (the flexibility of connexions between members) or essentially unknowable (the random small settlement of a column footing). Under these conditions it is meaningless to ask for a calculation of the 'actual' state of a structure; that state is an accidental product of the reaction between the structure and its environment, and can indeed change as a result of unpredictable events (a gale, an alteration in water table, a small earth tremor).

The prime structural problem remains the calculation of internal forces, in order that strength may be assessed, so that equilibrium equations, and a knowledge of the yield stress of the material, must certainly be retained. But elastic properties are introduced in a conventional analysis only for the purpose of solving the compatibility equations, and the strength of a ductile structure is independent of such faults as lack of fit of members, settlements of supports and so on. Boundary conditions must be considered as unknown, and there is then no need for elastic properties to enter the prime structural calculations. (Elastic properties will be needed to estimate, if required, overall deflexions of a structure.)

There remain, then, the equilibrium equations and the yield condition. These are the twin foundations of simple plastic theory; an equilibrium solution that satisfies yield is 'safe'. A conventional elastic design satisfies both conditions; the method may be uneconomical, and the predicted structural behaviour may be unobservable, but the method is also safe. A truly safe design, however, requires more than the application of simple plastic theory; it must be ensured that there is adequate 'ductility'. The structure as a whole, and its individual elements, must remain stable.

Bibliography

General histories

Ackermann, J.S. (1949). *Ars sine scientia nihil est*, Gothic theory of architecture at the Cathedral of Milan. *The Art Bulletin*, 84–111.

Benevento, E. (1991). *An introduction to the history of structural mechanics. Part I: Statics and resistance of solids. Part II: Vaulted structures and elastic systems*. 2 vols. Springer, New York.

Charlton, T.M. (1982). *A history of theory of structures in the nineteenth century*. Cambridge University Press.

Frankl, P. (1960). *The Gothic, literary sources and interpretations through eight centuries*. Princeton University Press.

Heyman, J. (1972). *Coulomb's memoir on statics*. Cambridge University Press. (Republished 1997, Imperial College Press, London.)

Oravas, G.AE. and McLean, L. (1966). Historical development of energetical principles in elasto mechanics. *Applied Mechanics Reviews*, **19**, no. 8, 647–58, no. 11, 919–33.

Saint-Venant, B. de (1864). Historique abrégé des recherches sur la résistance et sur l'élasticité des corps solides, pp. xc–cccxj in the 3rd edition of Navier's *Leçons*.

Straub, H. (1949). *Die Geschichte der Bauingenieurkunst*. Verlag Birkhaüser, Basle. (*A history of civil engineering*. Leonard Hill, London, 1952.)

Timoshenko, S.P. (1953). *History of strength of materials*. McGraw-Hill, New York and London.

Todhunter, I. and Pearson, K. (1886, 1893). *(A history of) The theory of elasticity (and of the strength of materials) (from Galilei to Lord Kelvin). Vol. I. Galilei to Saint-Venant 1639–1850. Vol. II. Saint-Venant to Lord Kelvin, Part I and Part II*. 2 volumes in 3 parts, Cambridge University Press. (Republished by Dover, New York, 1960.)

Truesdell, C. (1960). *The rational mechanics of flexible or elastic bodies 1638–1788*, Introduction to Leonhardi Euleri Opera Omnia, 2nd series, vol. XI(2). Orell Füssli, Zürich.

Secondary texts

Baker, J.F. (1954). *The steel skeleton, vol. 1, Elastic behaviour and design*. Cambridge University Press.

Baker, J.F., Horne, M.R. and Heyman, J. (1956). *The steel skeleton, vol. 2, Plastic behaviour and design*. Cambridge University Press.
Case, J. (1925). *The strength of materials*. Edward Arnold, London.
Case, J. and Chilver, A.H. (1959). *Strength of materials*. Edward Arnold, London.
Ewing, J.A. (1899). *The strength of materials*. Cambridge University Press.
Heyman, J. (1971). *Plastic design of frames, vol. 2, Applications*. Cambridge University Press.
Heyman, J. (1995). *The stone skeleton*. Cambridge University Press.
Joint Committee (of the Welding Institute and the Institution of Structural Engineers) (1964). *Fully-rigid multi-storey welded steel frames*. Supplement (1968); second report (1971). The Institution of Structural Engineers, London.
Neal, B.G. (1956). *The plastic methods of structural analysis*. Chapman and Hall, London (3rd edition 1977).
Pippard, A.J.S. and Baker, J.F. (1936). (4th edition 1968). *The analysis of engineering structures*. Edward Arnold, London.
Rankine, W.J.M. (1862). *A manual of civil engineering*. Griffin, London.
Shanley, F.R. (1957). *Strength of materials*. McGraw-Hill, New York.
Timoshenko, S. and Young, D.H. (1935). *Elements of strength of materials*. Van Nostrand, Princeton. (4th edition 1962).
Timoshenko, S. (1936). *Theory of elastic stability*. McGraw-Hill, New York.
Timoshenko, S.P. and Gere, J.M. (1961). *Theory of elastic stability*, 2nd edition. McGraw-Hill, New York.

Primary sources

Argyris, J.H. (1954, 1955). Energy theorems and structural analysis. *Aircraft Engineering*, **26**, 347–356, 383–389, 394; **27**, 42–58, 80–94, 125–134, 145–158.
Asimont, G. (1880). Hauptspannung und Sekundarspannung. *Zeitschrift für Baukunde*, **33**, 116.
Ayrton, W.E. and Perry, J. (1886). On struts. *The Engineer*, **62**, 464–5, 513–15.
Barlow, P. (1817). *An essay on the strength and stress of timber*. London.
Barlow, W.H. (1846). On the existence (practically) of the line of equal horizontal thrust in arches, and the mode of determining it by geometrical construction. *Minutes of Proceedings of the Institution of Civil Engineers*, **5**, 162–82.
Beggs, G.E. (1927). The use of models in the solution of indeterminate structures. *Journal of the Franklin Institute*, **203**, 375.
Bélidor, B.F. de (1729). *La science des ingénieurs dans la conduite des travaux de fortification et d'architecture civile*. Paris.
Bendixen, A. (1914). *Die Methode der Alpha-Gleichungen zur Berechnung von Rahmenkonstruktionen*. Berlin.
Bernoulli, Daniel (1741–43). De vibrationibus et sono laminarum elasticarum. *Commentarii Academiae Scientiarum Petropolitanae*, **13**, 105–120. Petersburg (1751).
Bernoulli, James (1691). Specimen alterum calculi differentialis. *Acta Eruditorum Lipsiae*. (In *Opera* (2 vols), no. XLII, vol. 1, p. 442, Geneva, 1744.)
Bernoulli, James (1694, 1695). Curvatura laminae elasticae. *Acta Eruditorum Lipsiae*, 262–276. (In *Opera*, vol. 1). Explicationes, annotationes et additiones. *Ibid.* 537–553. (In *Opera*, vol. 2).

Bernoulli, James (1705). Véritable hypothèse de la résistance des solides, avec la démonstration de la courbure des corps qui font ressort. *Histoire de l'Académie Royale des Sciences,* 176–186. Paris, 1706. (In *Opera* (2 vols), no. CII, vol. 2, 976–989.)

Bertot, J. (1855). [Contribution to discussion]. *Mémoires de la Société des Ingénieurs civils de France,* **8**, 278–280.

Betti, E. (1872). Teorema generale intorno alle deformazioni che fanno equilibrio a forze che agiscono soltanto alle superficie. *Il Nuovo Cimento,* ser. 2, **7**, 87 and **8**, 97.

Bleich, F. (1936). Calculation of statically indeterminate systems based on the theory of plasticity. *Preliminary Publication, International Association for Bridge and Structural Engineering, Second Congress,* 131–144. Berlin.

Bleich, H. (1932). Über die Bemessung statisch unbestimmter Stahl tragwerke unter Berücksichtigung des elastisch-plastischen Verhaltens des Baustoffes. *Bauingenieur,* **13**, 261.

Brown, E.H. (1967). Plastic asymmetric bending of beams. *International Journal of Mechanical Sciences,* **9**, 77–82.

Buffon, G.-L.L. (1740, 1741). Expériences sur la force du bois, *Histoire de l'Académie Royale des Sciences,* 543. Paris, 1742. Second Mémoire, *ibid.,* 292. Paris, 1744.

Bülffinger, G.B. (1729). De solidorum resistentia specimen. *Commentarii Academiae Scientiarum Petropolitanae,* **4**, 164–181 (1735).

Bullet, P. (1691). *L'architecture pratique.* Paris.

Castigliano, C.A.P. (1879). *Théorie de l'équilibre des systèmes élastiques et ses applications.* Turin. (Translated by Ewart S. Andrews: *Elastic stresses in structures,* Scott, Greenwood and Son, London, 1919. Republished with an introduction by G.AE. Oravas: *The theory of equilibrium of elastic systems and its applications,* Dover, New York, 1966.)

Charnes, A. and Greenberg, J.J. (1951). Plastic collapse and linear programming. *Bulletin of the American Mathematical Society,* **57**, 480.

Clapeyron, B.P.E. (1857). Calcul d'une poutre élastique reposant librement sur des appuis inégalement espacés. *Comptes Rendus hebdomadaires des Séances de l'Académie des Sciences,* **45**, 1076–80. Paris.

Clapeyron, B.P.E. (1858). Mémoires sue le travail des forces élastiques dans un corps solide élastique déformé par l'action des forces extérieures. *Comptes Rendus hebdomadaires des Séances de l'Académie des Sciences,* **46**, 208–212. Paris.

Clebsch, A. (1862). *Theorie der Elasticität fester Körper.* Leipzig. (Translated by Saint-Venant and Flamant, with notes by Saint-Venant: *Théorie de l'élasticité des corps solides.* Paris, 1883.)

Cotterill, J.H. (1865). On an extension of the dynamical principle of least action. *London, Edinburgh and Dublin Philosophical Magazine,* ser. 4, **29**, 299.

Coulomb, C.A. (1773). Essai sur une application des règles *de maximis & minimis* à quelques problèmes de statique, relatifs à l'architecture. *Mémoires de Mathématique & de Physique, présentés à l'Académie Royale des Sciences par divers Savans, & lûs dans ses Assemblés,* **7**, 343–82. Paris, 1776.

Couplet, P. (1729, 1730). De la poussée des voûtes. *Histoire de l'Académie Royale des Sciences,* 79, 117. Paris.

Cross, H. (1930). Analysis of continuous frames by distributing fixed-end moments. *Proceedings of the American Society of Civil Engineers,* **56**, 919–28.

Danyzy, A.A.H. (1732). Méthode générale pour déterminer la résistance qu'il
    faut opposer à la poussée des voûtes. *Histoire de la Société Royale des
    Sciences établie à Montpellier*, **2**, 40. Lyon, 1778.
Duleau, A. (1820). *Essai théorique et expérimental sur la résistance du fer forgé.*
    Paris.
Engesser, F. (1879). Über die Durchbiegung von Fachwerkträgern und die
    hierbei auftretenden zusätzlichen Spannungen. *Zeitschrift für Baukunde*, **29**,
    94–105.
Engesser, F. (1889). Über statisch unbestimmte Träger … *Zeitschrift des
    Architekten und Ingenieur-Vereins zu Hannover*, **35**, col. 733.
Engesser, F. (1892). *Die Zusatzkräfte und Nebenspannungen eiserner
    Fachwerkbrücken.* Berlin.
Euler, L. (1744). *Methodus inveniendi lineas curvas Maximi Minimive proprietate
    gaudentes, sive solutio problematis isoperimetrici latissimo sensu accepti.*
    Lausanne and Geneva.
Euler, L. (1757). Sur la force des colonnes. *Mémoires de l'Académie Royale des
    Sciences et Belles Lettres*, **13**, 252–282. Berlin.
Fairbairn, W. (1849). *An account of the construction of the Britannia and
    Conway tubular bridges.* London.
Feinberg, S.M. (1948). The principle of limiting stress (in Russian). *Prikladnaya
    Matematika i Mekhanika*, **12**, 63.
Frézier, A.F. (1737-39). *La théorie et la pratique de la coupe des pierres …*
    3 vols, Strasbourg and Paris.
Galilei, Galileo (1638). *Discorsi e Dimostrazioni Matematiche, intorno à due nuove
    scienze Attenenti alla Mecanica & i Movimenti Locali.* Elsevier, Leyden.
    (Facsimile reproduction, Brussels 1966.) (*Dialogues concerning two new
    sciences*, translated by Henry Crew and Alfonso de Salvio, Northwestern
    University Press, 1914, McGraw-Hill, New York, 1963; *Two new sciences*,
    translated by Stillman Drake, The University of Wisconsin Press, 1974.)
Gautier, H. (1717). *Dissertation sur l'épaisseur des culées des ponts …* Paris.
Girard, P.S. (An VI ≡ 1798). *Traité analytique de la résistance des solides, et des
    solides d'égale résistance.* Paris.
Girkmann, K. (1932). Über die Auswirkung der 'Selbsthilfe' des Baustahls in
    rahmenartigen Stabwerken. *Stahlbau*, **5**, 121.
Grandi, G. (1712). *La controversia contro dal Sig. A. Marchetti.* Lucca.
Grandi, G. (1712). *Risposta apologetica … alle opposizioni dal Sig. A. Marchetti.*
    Lucca.
Green, A.P. (1954). A theory of the plastic yielding due to bending of
    cantilevers and fixed-ended beams. *Journal of the Mechanics and Physics of
    Solids*, **3**, 1–15, 143–155.
Greenberg, H.J. and Prager, W. (1952). On limit design of beams and frames.
    Transactions of the American Society of Civil Engineers, **117**, 447.
Gregory, D. (1697). Catenaria. *Phil. Trans*, **19**, No. 231, 637–652.
Grüning, M. (1926). *Die Tragfähigkeit statisch unbestimmten Tragwerke aus Stahl
    bei beliebig häufig wiederholter Belastung.* Springer, Berlin.
Gvozdev, A.A. (1936). The determination of the value of the collapse load for
    statically indeterminate systems undergoing plastic deformation.
    Proceedings of the Conference on Plastic Deformations, December 1936,
    p. 19. Akademiia Nauk SSSR, Moscow-Leningrad (1938). (Translated by
    R.M. Haythornthwaite, *International Journal of Mechanical Sciences*, **1**,
    322–335, 1960.)

Hambly, E.C. (1985). Oil rigs dance to Newton's tune. *Proceedings of the Royal Institution of Great Britain*, **57**, 79–104.

Heyman, J. and Dutton, V.L. (1954). Plastic design of plate girders with unstiffened webs. *Welding and metal fabrication*, **22**, 265–272.

Heyman, J. (1970). The full plastic moment of an I-beam in the presence of shear force. *Journal of the Mechanics and Physics of Solids*, **18**, 359–365.

Heyman, J. (1976). Couplet's engineering memoirs 1726–33. *History of Technology*, ed. A. Rupert Hall and Norman Smith, 21–44. Mansell, London.

Hodgkinson, E. (1824). On the transverse strain, and strength of materials. *Memoirs of the Literary and Philosophical Society of Manchester*, 2nd series, **4**, 225.

Hodgkinson, E. (1831). Theoretical and experimental researches to ascertain the strength and best forms of iron beams. *Memoirs of the Literary and Philosophical Society of Manchester*, 2nd series, **5**, 407.

Hodgkinson, E. (1840) Experimental researches on the strength of pillars of cast iron and other materials. *Philosophical Transactions*, Part II, 385–486.

Hooke, Robert. (1675). *A description of helioscopes, and some other instruments.* London. (See Gunther, R.T. *Early science in Oxford*, **8**, 119–52. Oxford, 1931.)

Horne, M.R. (1950). Fundamental propositions in the plastic theory of structures. *Journal of the Institution of Civil Engineers*, **34**, 174.

Horne, M.R. (1951). The plastic theory of bending of mild steel beams with particular reference to the effect of shear forces. *Proc. Roy. Soc.* A, **207**, 216–228.

Horne, M.R. (1954). The flexural-torsional buckling of members of symmetrical I-section under combined thrust and unequal terminal moments. *Quarterly Journal of Mechanics and Applied Mathematics*, **7**, 410–426.

Jenkin, H.C. Fleeming. (1869). On the practical application of reciprocal figures to the calculation of strains on framework. *Transactions of the Royal Society of Edinburgh*, **25**, 441–7.

Jouravski, D.J. (1856). Sur la résistance d'un corps prismatique et d'une pièce composée en bois ou en tôle de fer à une force perpendiculaire à leur longeur. *Annales des ponts et chaussées, Mémoires*, **12**, 328–51.

Jouravski, D.J. (1860). Remarques sur les poutres en treillis et les poutres pleines en tôle. *Annales des Ponts et Chausées*, **20**, 113–134.

Kazinczy, G. (1914). Test with clamped beams. *Betonszemle*, **2**, 68–71, 83–87, 101–104 (in Hungarian). (See Kaliszky, S.: Gábor Kazinczy 1889–1964, *Periodica Polytechnica*, **28**, 75–93, Budapest, 1984, for an English summary of the 1914 paper.)

Kazinczy, G. (1931). Die Weiterentwicklung der Elastizitätstheorie. *Technika*. Budapest.

Kazinczy, G. (1936). Critical observations on the theory of plasticity. *Final Report, International Association for Bridge and Structural Engineering, Second Congress*, 56–69. Berlin (1939).

Kooharian, A. (1953). Limit analysis of voussoir (segmental) and concrete arches. *Proceedings of the American Concrete Institute*, **49**, 317.

Lagrange, J.L. (1770–73). Sur la figure des colonnes. *Miscellanea Taurinensia*, **5**, 123–166. Turin.

La Hire, P. de (1695). *Traité de Mécanique*. Paris.

La Hire, P. de (1712). Sur la construction des voûtes dans les édifices. *Mémoires de l'Académie Royale des Sciences,* **69.** Paris (1731).

Lamarle, E. (1845, 1846). Mémoire sur la flexion du bois. *Annales des Travaux publics de Belgique,* III, 1–64 and IV, 1–36.

Lamé, G. (1852). *Leçons sur la théorie mathématique de l'élasticité des corps solides.* Paris.

Leibniz, G.W. (1684). Demonstrationes novae de Resistentiâ solidorum. *Acta Eruditorum Lipsiae,* 319–325.

Leibniz, G.W. (1691). De solutionibus problematis catenarii … *Acta Eruditorum,* 435–439.

Le Seur, T., Jacquier, F. and Boscovich, R.G. (1743). *Parere di tre mattematici sopra i danni che si sono trovati nella cupola di S. Pietro.* Rome.

Leth, C.-F.A. (1954). Effect of shear stresses on the carrying capacity of I-beams. *Technical Report No. A-11-107,* Brown University.

Livesley, R.K. (1964). *Matrix methods of structural analysis.* Pergamon Press, Oxford. (2nd edition, 1975).

Lowe, P.G. (1971). *Classical theory of structures.* Cambridge University Press.

Macaulay, W.H. (1919). A note on the deflection of beams. *Messenger of mathematics,* **48,** 129.

Maier-Leibnitz, H. (1936). Test results, their interpretation and application. *Preliminary Publication, International Association for Bridge and Structural Engineering, Second Congress,* 97–130. Berlin.

Manderla, H. (1880). Die Berechnung der Sekundärspannungen. *Allgemeine Bauzeitung,* **45,** 34.

Marchetti, A. (1669). *De resistentia solidorum.* Florence.

Mariotte, E. (1686). *Traité du mouvement des eaux.* Paris.

Maxwell, J.C. (1864). On the calculation of the equilibrium and stiffness of frames. *London, Edinburgh and Dublin Philosophical Magazine,* ser. 4, **27,** 294.

Melan, E. (1936). Theory of statically indeterminate systems. *Preliminary Publication, International Association for Bridge and Structural Engineering, Second Congress,* 43–64. Berlin.

Michell, A.G.M. (1899). Elastic stability of long beams under transverse forces. *Philosophical Magazine,* **48,** 298–309.

Mohr, O.C. (1874). Beitrag zur Theorie der Bogenfachwerksträger. *Zeitschrift des Architekten und Ingenieur-Vereins zu Hannover,* **20,** col. 223.

Mohr, O.C. (1892, 1893). Die Berechnung der Fachwerk mit starren Kustenverbindungen. *Der Civilingenieur,* **38,** 577, and **39,** 67.

Müller-Breslau, H.F.B. (1883). Zur Theorie der versteifung labiler flexibiler Bogenträger. *Zeitschrift für Bauwesen,* **33,** 312.

Müller-Breslau, H.F.B. (1886). Zur Theorie der Biegungsspannungen in Fachwerksträgern. *Zeitschrift des Architekten und Ingeneiur-Vereins zu Hannover,* **32,** col. 399.

Musschenbroek, P. van (1729). *Physicae experimentales, et geometricae, …Dissertationes.* Leyden.

Navier, C.L.M.H. (1826). *Résumé des leçons données à l'École des Ponts et Chaussées, sur l'application de la mécanique à l'établissement des constructions et des machines.* Paris (2nd edition, 1833; 3rd edition, with notes and appendices by B. de Saint-Venant, 1864.)

Neal, B.G. (1951). The behaviour of framed structures under repeated loading. *Quarterly Journal of Mechanics and Applied Mathematics,* **4,** 78.

Neal, B.G. (1961*a*). The effect of shear and normal forces on the fully plastic moment of a beam of rectangular cross section. *Journal of Applied Mechanics*, **28**, 269–274.

Neal, B.G. (1961*b*). Effect of shear force on the fully plastic moment of an I-beam. *Journal of mechanical engineering science*, **3**, 258–266.

Neal, B.G. and Symonds, P.S. (1950–51). The calculation of collapse loads for framed structures. *Journal of the Institution of Civil Engineers*, **35**, 21.

Neal, B.G. and Symonds, P.S. (1952). The rapid calculation of the plastic collapse load for a framed structure. *Proceedings of the Institution of Civil Engineers*, **1** (Part 3), 58.

Ogle, M.H. (1964). *Shakedown of steel frames*. Ph.D. Thesis, Cambridge University.

Pain, J.F. and Roberts, Gilbert (1933–34). Sydney Harbour Bridge: calculations for the steel superstructure. *Minutes of Proceedings of the Institution of Civil Engineers*, **238**, 256–309.

Pardies, I.G. (1673). *La statique ou la science des forces mouvantes*. Paris. (2nd edition 1674.)

Parent, A. (1713). *Essais et recherches de Mathématique et de Physique*. 3 vols. Paris.

Persy, N. (1834). *Cours de stabilité des constructions à l'usage des élèves de l'École d'application de l'artillerie et du génie*. Metz. (Lithographed notes.)

Pippard, A.J.S., Tranter, E. and Chitty, L. (1936). The mechanics of the voussoir arch. *Journal of the Institution of Civil Engineers*, **4**, 281.

Pippard, A.J.S. and Ashby, R.J. (1938). An experimental study of the voussoir arch. *Journal of the Institution of Civil Engineers*, **10**, 383.

Pitot, H. (1726). Examen de la force qu'il faut donner aux cintres dont on se sert dans la construction des grandes voûtes des arches des ponts, &c. *Mémoires de l'Académie Royale des Sciences*, 216. Paris.

Poleni, G. (1748). *Memoire istoriche della gran Cupola del Tempio Vaticano*. Padova.

Prandtl, L. (1899). *Kipperscheinungen*. (Dissertation). Munich.

Prony, R. de (An X ≡ 1802). *Recherches sur la poussée des terres...*, Paris.

Robertson, A. and Cook, G. (1913). Transition from the elastic to the plastic state in mild steel. *Proc. Roy. Soc. A*, **88**.

Robertson, A. (1925). The strength of struts. *Selected Engineering Papers*, No. 28. The Institution of Civil Engineers, London.

Saint-Venant, B. de (1855). Mémoire sur la torsion des prismes ..., *Comptes rendues de l'Académie des Sciences, Mémoires des Savants étrangers*, **14**, 233–560. Paris.

Saint-Venant, B. de (1856). Mémoire sur la flexion des prismes ..., *Journal de mathématiques de Liouville*, 2nd series, **1**, 89–189.

Southwell, R.V. (1940). *Relaxation methods in engineering science*. Oxford.

Steel Structures Research Committee, *First Report*, 1931, *Second Report*, 1934, *Final Report*, 1936. HMSO, London.

Stirling, J. (1717). *Lineae Tortii Ordinis Neutonianae*. Oxford.

Symonds, P.S. and Prager, W. (1950). Elastic-plastic analysis of structures subjected to loads varying arbitrarily between prescribed limits. *Journal of Applied Mechanics*, **17**, 315.

Timoshenko, S.P. (1905, 1906). Lateral buckling of beams. *Bull. Polytech. Inst. St Petersburg*, **4, 5** (in Russian).

Timoshenko, S.P. (1910). Einige Stabilitätsprobleme der Elastizitätstheorie. *Z. Math. u. Physik*, **58**, 337–385.

Tredgold, T. (1822). *A practical essay on the strength of cast iron*. London.

Varignon, P. (1702). De la résistance des solides ... *Histoire de l'Académie Royale des Sciences*, 66-94. Paris, 1704.

Vauban, Marquis de (1704, 1706). *Traité de l'attaque des places. Traité de la défense des places*. Paris.

Vitruvius (c. 30 BC). *Architectura*. (Many translations.)

Vlasov, V.Z. (1940). *Thin walled elastic beams*. (Translated from the Russian, Israel Program for Scientific Translations, Jerusalem, 1961.)

Ware, S. (1809). *A treatise of the properties of arches, and their abutment piers*. London.

Winkler, E. (1881). *Vorträge über Brückenbau, Theorie der Brücken*. 2 vols. Vienna.

Winkler, E. (1881). Die Sekundärspannungen in Eisenkonstruktionen. *Deutsche Bauzeitung*, **46**, 72–9.

Wittrick, W.H. (1965). A generalization of Macaulay's method with applications in structural mechanics. *American Institute of Aeronautics and Astronautics Journal*, **3**, 326.

Yvon Villarceau, A. (1854). L'établissement des arches de pont. *Institut de France, Académie des Sciences, Mémoires présentés par divers savants*, **12**, 503.

Name Index

171

Subject Index